Asking Questions in Biology

We work with leading authors to develop the
strongest educational materials in biology,
bringing cutting-edge thinking and best learning
practice to a global market.

Under a range of well-known imprints, including
Prentice Hall, we craft high quality
print and electronic publications which help
readers to understand and apply their content,
whether studying or at work.

To find out more about the complete range of our
publishing please visit us on the World Wide Web at:
www.pearsoned.co.uk

Asking Questions in Biology

Key Skills for Practical Assessments and Project Work

Second Edition

Chris Barnard, Francis Gilbert

School of Life and Environmental Sciences, University of Nottingham

and Peter McGregor

Zoological Institute, University of Copenhagen

Harlow, England • London • New York • Boston • San Francisco • Toronto
Sydney • Tokyo • Singapore • Hong Kong • Seoul • Taipei • New Delhi
Cape Town • Madrid • Mexico City • Amsterdam • Munich • Paris • Milan

Pearson Education Limited
Edinburgh Gate
Harlow
Essex CM20 2JE
England

and Associated Companies throughout the world

Visit us on the World Wide Web at:
www.pearsoned.co.uk

First published under the Longman imprint 1993
Second edition 2001

© Pearson Education Limited 1993, 2001

ISBN 0 130 90370 1

British Library Cataloguing-in-Publication Data
A catalogue record for this book is available from the British Library

Library of Congress Cataloguing-in-Publication Data
Barnard, C. J. (Christopher J.)
    Asking questions in biology : key skills for practical assessments and project
    work / Chris Barnard, Francis Gilbert and Peter McGregor.–2nd ed.
        p. cm.
    Includes bibliographical references (p.).
    ISBN 0–13–090370–1
    1. Science–Methodology. 2. Biology–Methodology. I. Gilbert, Francis S.
(Francis Sylvest), 1956– II. McGregor, Peter K. III. Title.

Q175 .B218 2001
570′.7′2–dc21   2001021820

10 9 8 7 6 5 4 3
07 06 05 04

Typset in 9.5/12pt Concorde BE by 35
Printed in Great Britain by Henry Ling Limited at the Dorset Press,
Dorchester, DT1 1HD

Contents

Preface

Science is a process of asking questions, in most cases precise, quantitative questions that allow distinctions to be drawn between alternative explanations of events. Asking the right questions in the right way is a fundamental skill in scientific enquiry, yet in itself it receives surprisingly little explicit attention in scientific training. Students being trained in scientific subjects, for instance in sixth forms, colleges and universities, learn the factual science and some of the tools of enquiry such as laboratory techniques, mathematics, statistics and computing, but they are taught little about the process of question-asking itself.

The first edition of this book had its origins in a first-year practical course we, and now others, have run at the University of Nottingham for several years. The aim of the course is to introduce students in the biological sciences to the skills of observation and enquiry, but focusing on the process of enquiry – how to formulate hypotheses and predictions from raw information, how to design critical observations and experiments, and how to choose appropriate analyses – rather than on laboratory, field and analytical techniques *per se*. This focus is maintained in the second edition. However, we have also responded to a number of positive suggestions that people have been kind enough to make since the book first appeared and which we think enhance its usefulness in teaching practical biology generally.

The first major change is to the four main examples that work their way through the book. We have broadened these to include a range of subjects – parasitology, plant ecology and toxicology – as well as the behaviour of the first edition, so that the wider applicability of the principles we discuss are clearer. In choosing these, we have been mindful of the diversity of approaches in practical teaching, from set-piece laboratory exercises to open-ended projects and field courses where these skills become important. Some of the new examples therefore have greater scope, and could perhaps lead to more ambitious investigations, than the original behavioural examples.

The second major change is to introduce more discussion of parametric statistics and their assumptions. Parametric tests are the first port of call for many teachers and students, and we felt it was important to broaden their coverage accordingly. We have thus incorporated parametric equivalents to each of the nonparametric tests covered previously, and discuss more fully the criteria for choosing one family of tests over the other. Some of the parametric tests are more complex and sophisticated than their nonparametric counterparts, and one's first choice would undoubtedly be to run them on a suitable computer package rather than work through them by hand. Nevertheless, full worked examples are provided for all the tests covered. As before, statistics are introduced as aids at the appropriate stage of enquiry, and we emphasize their role and the information they provide rather than the theory and mechanics behind them. The book, like its earlier edition, is not intended to be a statistics text.

As well as these changes, we have introduced numerous additional topics, such as: power tests, *post hoc* analysis and the use of covariates; added more follow-up references; and produced a comprehensive new Quick Test Finder.

The book looks at the process of enquiry during its various stages, starting with unstructured observations of a range of material and working through to the production of a complete written report. In each chapter, different skills are emphasized and the main examples running through the book illustrate their application at each stage.

The book looks at the process of enquiry during its various stages, starting with unstructured observations and working through to the production of a complete written report. In each section, different skills are emphasized and a series of main examples runs through the book to illustrate their application in each stage.

The book begins with a look at scientific question-asking in general. How do we arrive at the right questions to ask? What do we have to know before we can ask sensible questions? How should questions be formulated to be answered most usefully? Chapter 1 addresses these points by looking at the development of testable hypotheses and predictions and the sources from which they might arise.

Chapter 2 looks at how hypotheses and predictions can be derived from unstructured observational notes. Exploratory analysis is an important first step in deriving hypotheses from raw data and the chapter introduces plots and summary statistics as useful ways of identifying interesting patterns on which to base hypotheses. The chapter concludes by pointing out that although hypotheses and their predictions are naturally specific to the investigation in hand, testable predictions in general fall into two distinct groups: those dealing with some kind of *difference* between groups of data and those dealing with a *trend* in the quantitative relationship between groups of data.

The distinction between difference and trend predictions is developed further in Chapter 3, which discusses the use of confirmatory analyses. The concept of statistical significance is introduced as an arbitrary but generally agreed yardstick as to whether observed differences or trends are interesting, and a number of basic but broadly applicable significance tests are explained. Throughout, however, the emphasis is on the use of such tests as tools of enquiry rather than on the statistical theory underlying them. Having introduced significance tests and some potential pitfalls in their use, the book uses the main worked examples to show how some of their predictions can be tested and hypotheses refined in the light of testing.

In Chapter 4, the book considers the presentation of information. Once hypotheses have been tested, how should the outcome be conveyed for greatest effect? The chapter discusses the use of tables, figures and other modes of presentation and shows how a written report should be structured. The points made in the chapter are then illustrated in a complete written report based on one of the main worked examples.

At the end of the book are a number of appendices. These provide some self-test questions and answers based on the material in the book, some worked examples of significance tests and some statistical tables for use in significance testing.

We said that the book had its inception in our introductory practical course. However, the practical course itself was developed in response to an increasingly voiced need on the part of students to be taught how to ask and answer questions. Both the practical course and the book have benefited immensely from a constant and pleasurable interaction with our undergraduates over the years. Their enquiries and insights continue to hone the way we teach and have been the guiding force behind all the discussions in the book.

Finally, we should like thank all the people who have commented on the book since its first appearance and encouraged us to think about the amendments we have made in this present edition. Two anonymous reviewers of the draft of the second edition made especially helpful suggestions and the final product was materially improved by their inclusion.

Chris J. Barnard
Francis S. Gilbert
Peter K. McGregor
October, 2000

Acknowledgements

We are grateful to Blackwell Science Ltd for permission to reproduce copyright material: Figures 2.6 and 3.3 modified from *Choosing and using statistics*, published by Blackwell Science Ltd, (Dytham, C. 1999).

Companion Web Site

Material will be available on the internet to support students and lecturers using this book, at the web address *www.booksites.net/barnard*. This will comprise downloadable Microsoft Excel® sheets that will enable the student or lecturer to carry out in Excel all of the statistical tests described and included within the book. The tests covered will include χ^2, correlation and regression, and the specific and general hypotheses tested by analyses of variance, both parametric and non-parametric. The package will form a useful supporting resource to the text.

CHAPTER 1

Doing science

Where do questions come from?

You're out for a walk one autumn afternoon when you notice a squirrel picking up acorns under some trees. Several things strike you about the squirrel's behaviour. For one thing it doesn't seem to pick up all the acorns it comes across; a sizeable proportion is ignored. Of those it does pick up, only some are eaten. Others are carried up into a tree where the squirrel disappears from view for a few minutes before returning to the supply for more. Something else strikes you: the squirrel doesn't carry its acorns up the nearest tree but instead runs to one several metres away. You begin to wonder why the squirrel behaves in this way. Several possibilities occur to you. Although the acorns on the ground all look very similar to you, you speculate that some might contain more food than others, or perhaps they are easier to crack. By selecting these, the squirrel might obtain food more quickly than by taking indiscriminately any acorn it encountered. Similarly, the fact that it appears to carry acorns into a particular tree suggests this tree might provide a more secure site for storing them.

While all these might be purely casual reflections, they are revealing of the way we analyse and interpret events around us. The speculations about the squirrel's behaviour may seem clutched out of the air on a whim but they are in fact structured around some clearly identifiable assumptions, for instance that achieving a high rate of food intake matters in some way to the squirrel and influences its preferences, and that using the most secure storage site is more important to it than using the most convenient site. If you wanted to pursue your curiosity further, these assumptions would be critical to the questions you asked and the investigations you undertook. If all this sounds very familiar to you as a science student it should, because, whether you intended it or not, your speculations are essentially scientific. Science is simply formalized speculation backed up (or otherwise) by equally formalized observation and experimentation. In its broadest sense most of us 'do science' all the time.

1.1 Science as asking questions

Science is often regarded by those outside it as an open-ended quest for objective understanding of the universe and all that is in it. But this is so only in a rather trivial sense. The issue of objectivity is a thorny one and, happily, well beyond the scope of this book. Nevertheless, the very real constraints that limit human objectivity mean that use of the term must at least be hedged about with serious qualifications. The issue of open-endedness is really the one that concerns us here. Science is open-ended only in that its directions of enquiry are, in principle, limitless. Along each path of enquiry, however, science is far from open-ended. Each step on the way is, or should be, the result of refined question-asking, a narrowing down of questions and methods of answering them to provide the clearest possible distinction between alternative explanations for the phenomenon in hand. This is a skill, or series of skills really, that has to be acquired, and acquiring it is one of the chief objectives of any scientific training.

While few scientists would disagree with this, identifying the different skills and understanding how training techniques develop them are a lot less straightforward. With increasing pressure on science courses in universities and colleges to teach more material to more people and to draw on an expanding and increasingly sophisticated body of knowledge, it is more important than ever to understand how to marshal information and direct enquiry. This book is the result of our experiences in teaching investigative skills to university undergraduates in the life sciences. It deals with all aspects of scientific investigation, from thinking up ideas and making initial exploratory observations, through developing and testing hypotheses, to interpreting results and preparing written reports. It is not an introduction to data-handling techniques or statistics, although it includes a substantial element of both; it simply introduces these as tools to aid investigation. The theory and mechanics of statistical analysis can be dealt with more appropriately elsewhere.

The principles covered in the book are extraordinarily simple, yet, paradoxically, students find them very difficult to put into practice when taught in a piecemeal way across a number of different courses.

The book has evolved out of our attempts to get over this problem by using open-ended, self-driven practical exercises in which the stages of enquiry develop logically through the desire of students to satisfy their own curiosity. However, the skills it emphasizes are just as appropriate to more limited set-piece practicals. Perhaps a distinction – admittedly over-generalized – that could be made here, and which to some extent underpins our preference for a self-driven approach, is that with many set-piece practicals it is obvious *what* one is supposed to do but often not *why* one is supposed to do it. Almost the opposite is true of the self-driven approach; here it is clear why any investigation needs to be undertaken but usually less clear what should be done to see it through successfully. In our experience, developing the 'what' in the context of a clear 'why' is considerably more instructive than attempting to reconstruct the 'why' from the 'what' or, worse, ignoring it altogether.

1.2   Basic considerations

Scientific enquiry is not just a matter of asking questions; it is a matter of asking the *right questions* in the *right way*. This is more demanding than it sounds. For a start, it requires that something is known about the system or material in which an investigator is interested. A study of mating behaviour in guppies, for instance, demands that you can at least tell males from females and recognize courtship and copulation. Similarly, it is difficult to make a constructive assessment of parasitic worm burdens in host organisms if you are ignorant of likely sites of infection and can't tell worm eggs from faecal material.

Of course, there are several ways in which such knowledge can be acquired: e.g. textbooks, specialist journals, asking an expert or simply finding out for yourself through observation and exploration. Whichever way is most appropriate, however, a certain amount of background preparation is usually essential, even for the simplest investigations. In practical classes, some background is usually provided for you in the form of handouts or accompanying lectures, but the very variability of biological material means that generalized and often highly stylized summaries are poor substitutes for hard personal experience. Nevertheless, given the inevitable constraints of time, materials and available expertise, they are usually a necessary second best. There is also a second, more important, reason why there is really no substitute for personal experience: the information you require may not exist or, if it does exist, it may not be correct. Taking received wisdom at face value can be a dangerous business – something even seasoned researchers can continue to discover, the famous geneticist and biostatistician R. A. Fisher among them.

In the early 1960s, Fisher and other leading authorities at the time were greatly impressed by an apparent relationship between duodenal ulcer and certain rhesus and MN blood groups. Much intellectual energy was expended trying to explain the relationship. A sceptic, however, mentioned the debate to one of his blood-group technicians. The technician, for years at the sharp end of blood-group analysis, resolved the issue on the spot. The relationship was an artefact of blood transfusion! Patients with ulcers had received transfusions because of haemorrhage. As a result, they had temporarily picked up rhesus and MN antigens from their donors. When patients who had not been given transfusions were tested, the relationship disappeared (Clarke, 1990).

Where at all feasible, therefore, testing assumptions yourself and making up your own mind about the facts available to you is a good idea. It is impossible to draw up a definitive list of what it is an investigator needs to know as essential background; biology is too diverse a subject and every investigation is to some extent unique in its factual requirements. Nevertheless, it is useful to indicate the kinds of information that are likely to be important. Some examples might be as follows:

Question   *Can the material of interest be studied usefully under laboratory conditions or will unavoidable constraints or manipulations so affect it that any conclusions will have only dubious relevance to its normal state or functions?*

For instance, can mating preferences in guppies usefully be studied in a small plastic aquarium, or will the inevitable restriction on movement and the impoverished environment compromise normal courtship activity?

Or, if nutrient transfer within a plant can be monitored only with the aid of a vital dye, will normal function be maintained in the dyed state or will the dye interfere subtly with the processes of interest?

Question   *Is the material at the appropriate stage of life history or development for the desired investigation?*

There would, for instance, be little point in carrying out vaginal smears on female mice to establish stages of the oestrous cycle if some females were less than 28 days of age. Such mice may well not have begun cycling.

Likewise, it would be fruitless to monitor the faeces of infected mice for the eggs of a nematode worm until a sufficient number of days have passed after infection for the worms to have matured.

Will the act of recording from the material affect its performance?

For example, removing a spermatophore (package of sperm donated by the male) from a recently mated female cricket in order to assay its sperm content may adversely affect the female's response to males in the future.

Or, the introduction of an intracellular probe might disrupt the aspect of cell physiology it was intended to record.

Has the material been prepared properly?

If the problem to be investigated involves a foraging task (e.g. learning to find cryptic prey), has the subject been trained to perform in the apparatus and has it been deprived of food for a short while to make it hungry?

Similarly, if a mouse of strain X is to be infected with a particular blood parasite so that the course of infection can be monitored, has the parasite been passaged in the strain long enough to ensure its establishment and survival in the experiment?

Does the investigation make demands on the material that it is not capable of meeting?

Testing for the effects of acclimation on some measure of coping in a new environment might be compromised if conditions in the new

environment are beyond those the organism's physiology or behaviour have evolved to meet.

Likewise, testing a compound from an animal's environment for carcinogenic properties in order to assess risk might not mean much if the compound is administered in concentrations or via routes that the animal could never experience naturally.

Question   *Are assumptions about the material justified?*

In an investigation of mating behaviour in dragonflies, we might consider using the length of time a male and female remain coupled as an index of the amount of sperm transferred by the male. Before accepting this, however, it would be wise to conduct some pilot studies to make sure it was true; it might be, for instance, that some of the time spent coupled reflected mate-guarding rather than insemination.

By the same token, assumptions about the relationship between the staining characteristics of cells in histological sections and their physiological properties might need verifying before concluding anything about the distribution of physiological processes within an organ.

The list could go on for a long time, but these examples are basic questions of practicality. They are not very interesting in themselves but they, and others like them, need to be addressed before interesting questions can be asked. Failure to consider them will almost inevitably result in wasted time and materials.

Of course, even at this level, the investigator will usually have the questions ultimately to be addressed – the whole point of the investigation – in mind, and these will naturally influence initial considerations.

1.3 The skill of asking questions

1.3.1 Testing hypotheses

Charles Darwin once remarked that without a hypothesis a geologist might as well go into a gravel pit and count the stones. He meant, of course, that simply gathering facts for their own sake was likely to be a waste of time. A geologist is unlikely to profit much from knowing the number of stones in a gravel pit. This seems self-evident, but such *undirected* fact-gathering (not to be confused with the often essential descriptive phase of hypothesis development) is a common problem among students in practical and project work. There can't be many science teachers who have not been confronted by a puzzled student with the plea: 'I've collected all these data, now what do I do with them?' The answer, obviously, is that the investigator should know what is to be done with the data before they are collected. As Darwin well knew, what gives data collection direction is a working *hypothesis*.

The word 'hypothesis' sounds rather formal and, indeed, in some cases hypotheses may be set out in a tightly constructed, formal way. In more general usage, however, its meaning is a good deal looser. Verma and Beard (1981), for example, define it as simply

> a tentative proposition which is subject to verification through sub-sequent investigation. . . . In many cases hypotheses are hunches that the researcher has about the existence of relationships between variables

A hypothesis, then, can be little more than an intuitive feeling about how something works, or how changes in one factor will relate to changes in another, or about any aspect of the material of interest. However vague it may be, though, it is formative in the purpose and design of investigations because these set out to test it. If at the end of the day the results of the investigation are at odds with the hypothesis, the hypothesis may be rejected and a new one put in its place. As we shall see later, *hypotheses are never proven, merely rejected if data from tests so dictate, or retained for the time being for possible rejection after further tests.*

1.3.2   How is a hypothesis tested?

If a hypothesis is correct, certain things will follow. Thus if our hypothesis is that a particular visual display by a male chaffinch is sexual in motivation, we might expect the male to be more likely to perform the display when a female is present. Hypotheses thus generate *predictions*, the testing of which increases or decreases our faith in them. If our male chaffinch turned out to display mainly when other males were around and almost never with females, we might want to think again about our sexual motivation hypothesis. However, we should be wrong to dismiss it solely on these grounds. It could be that such displays are important in defending a good quality breeding territory which eventually will attract a female. The context of the display could thus still be sexual, but in a less direct sense than we had

first considered. In this way, hypotheses can produce tiers of more and more refined predictions before they are rejected or tentatively accepted. Making such predictions is a skilled business because each must be phrased so that testing it allows the investigator to discriminate in favour of or against the hypothesis. While it is best to phrase predictions as just that (thus: *males will perform more of display* y *in the presence of females*), they sometimes take the form of questions (*do males perform more of display* y *when females are present?*). The danger with the question format, however, is that it can easily become too woolly and vague to provide a rigorous test of the hypothesis (e.g. *do males behave differently when females are present?*). Having to phrase a precise prediction helps counteract the temptation to drift into vagueness.

Hypotheses, too, can be so broad or imprecise that they are difficult to reject. In general the more specific, mutually exclusive hypotheses that can be formulated to account for an observation the better. In our chaffinch example, the first hypothesis was that the display was sexual. Another might be that it reflected aggressive defence of food. Yet another that it was an anti-predator display. These three hypotheses give rise to very different predictions about the behaviour and it is thus, in principle, easy to distinguish between them. As we have already seen, however, distinguishing between the 'sexual' and 'aggressive' hypotheses may need more careful consideration than we first expect. *Straw man hypotheses* are another common problem. Unless some effort has gone into understanding the material, there is a risk of setting up hypotheses that are completely inappropriate. Thus, suggesting that our displaying chaffinch was demonstrating its freedom from avian malaria would make little sense in an area where malaria was not endemic. We shall look at the development of hypotheses and their predictions in more detail later on.

1.4   Where do questions come from?

As we have already intimated, questions do not spring out of a vacuum. They are triggered by something. They may arise from a number of sources.

1.4.1   Curiosity

Questions arise naturally when thinking about any kind of problem. Simple curiosity about how something works or why one group of organisms differs in some way from another group can give rise to useful questions from which testable hypotheses and their predictions can be derived. There is nothing wrong with 'armchair theorizing' and 'thought experiments' as long as, where possible, they are put to the test. Sitting in the bath and wondering about how migratory birds manage to navigate, for example, could suggest roles for various environmental cues like the sun, stars and topographical features. This in turn could lead to hypotheses about how they are used and predictions about the effects of removing or altering them. By the time the water was cold, some useful preliminary experiments might even have been devised.

1.4.2   Casual observation

Instead of dreaming in the bath, you might be watching a tank full of fish, or sifting through some histological preparations under a microscope. Various things might strike you. Some fish in the tank might seem very aggressive, especially towards others of their own species, but this aggressiveness might occur only around certain objects in the tank, perhaps an overturned flowerpot or a clump of weed. Similarly, certain cells in the histological preparations may show unexpected differences in staining or structure. Even though these aspects of fish behaviour and cell appearance were not the original reason for watching the fish or looking at the slides, they might suggest interesting hypotheses for testing later. A plausible hypothesis to account for the behaviour of the fish, for instance, is that the localized aggression reflects territorial defence. Two predictions might then be: (a) *on average, territory defenders will be bigger than intruders* (because bigger fish are more likely to win in disputes and thus obtain a territory in the first place) and (b) *removing defendable resources like upturned flowerpots will lead to a reduction in aggressive interactions*. Similarly, a hypothesis for differences in cell staining and structure is that they are due to differences in the age and development of the cells in question. A prediction might then be: *younger tissue will con-tain a greater proportion of* (what are conjectured to be the) *immature cell types.*

1.4.3   Exploratory observations

It may be that you already have a hypothesis in mind, say that a par-ticular species of fish will be territorial when placed in an appropriate aquarium environment. What is needed is to decide what an appro-priate aquarium environment might be so that suitable predictions can be made to test the hypothesis. Obvious things to do would be to play around with the size and number of shelters, the position and quality of feeding sites, the number and sex ratio of fish introduced into the tank and so on. While the effects of these and other factors on territorial aggressiveness among the fish might not have been guessed at beforehand, such manipulations are likely to suggest rela-tionships with aggressiveness which can then be used to predict the outcome of further, *independent* investigations. Thus if exploratory results suggested aggressiveness among defending fish was greater when there were ten fish in the tank compared with when there were five, it would be reasonable to predict that aggressiveness would increase as the number of fish increased, *all other things being equal.* An experiment could then be designed in which shelters and feeding sites were kept constant but different numbers of fish, say two, four, six, eight, ten or 15, were placed in the tank. Measuring the amount of aggression by a defender with each number of fish would provide a test of the prediction.

1.4.4   Previous studies

One of the richest sources of questions is, of course, past and ongoing research. This might be encountered either as published literature or 'live' as research talks at conferences or seminars. A careful reading of most published papers, articles or books will turn up ideas for fur-ther work, whether at the level of alternative hypotheses to explain the problem in hand or at the level of further or more discriminating predictions to test the current hypothesis. Indeed, this is the way most of the scientific literature develops. Some papers, often in the form of mathematical models or speculative reviews, are specifically intended to generate hypotheses and predictions and may make no attempt to test them themselves. At times, certain research areas can become overburdened with hypotheses and predictions, generating more than people are able or have the inclination to test. If this happens, it can have a paralysing effect on the development of research. It is thus important that hypotheses, predictions and tests proceed as nearly as is feasible hand in hand.

1.5   What this book is about

We've said a little about how science works and how the kind of question-asking on which it is based can arise. We now need to look at each part of the process in detail because while each may seem straightforward in principle, some knotty problems can arise when science is put into practice. In what follows, we shall see how to:

1. frame hypotheses and predictions from preliminary source material;

2. design experiments and observations to test predictions;

3. analyse the results of tests to see whether they confirm our original hypothesis; and

4. present the results and conclusions of tests so that they are clear and informative.

The discussion deals with these aspects in order so that the book can be read straight through or dipped into for particular points. A summary at the end of each chapter highlights the important take-home

messages and the self-test questions at the end show what you should be able to tackle after reading the book.

Remember, the book is about asking and answering questions in biology – it is not a biology textbook or a statistics manual and none of the points it makes are restricted to the examples that illustrate them. At every stage you should be asking yourself how what it says might apply in other biological contexts, especially if you have an interest in investigating them!

References

Clarke, C. (1990) Professor Sir Ronald Fisher FRS. *British Medical Journal* **301**, 1446–1448.

Verma, G. K. and Beard, R. M. (1981) *What is educational research? Perspectives on techniques of research*. Gower, Aldershot.

CHAPTER 2

Asking questions

The art of framing hypotheses and predictions

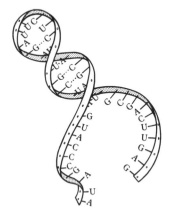

So far, we've discussed asking questions in a very general way. Simply being told that science proceeds like this, however, is not particularly helpful unless it is clear how the principles can be applied to the situation in hand. The idea of this chapter is thus to look at the development of the procedure in the context of various investigations that you might undertake in practical and project work. We shall assume for the moment that the material of interest is derived from your own observations. We shall start, therefore, with the problem of making observations and directing them in order to produce useful information.

2.1 Observation

2.1.1 Observational notes and measurements

When first confronted with an unfamiliar system, it is often difficult to discern anything of interest straight away. This seems to be true regardless of the complexity of the system. For instance, a common cause of early despair among students watching animals in a tank or arena for the first time is the mêlée of ceaselessly changing activities, many of which seem directionless and without obvious goal. An equally common complaint is that the animals seem to be doing nothing at all worth mentioning. It is not unusual for *both* extremes to be generated by the same animals.

In both of the above cases, the problem almost always turns out to be not what the animals are or are not doing, but the ability of the observer to observe. This is because observation involves more than just staring passively at material on the assumption that if anything about it is interesting then it will also be obvious. To be revealing, observations may need to be very systematic, perhaps involving manipulations of the material to see what happens. They may involve measurements of some kind since some things may be apparent quantitatively rather than qualitatively. In themselves, therefore, observations are likely to involve a certain amount of question-asking. Their ultimate purpose, however, is to provide the wherewithal to frame testable hypotheses and the discriminating predictions that will distinguish between the hypotheses.

To see how the process works, we shall first give some examples of observational notes and then look at the way these can be used to derive hypotheses and predictions. These examples, therefore, develop through the book from initial observational notes, through framing and testing hypotheses, to producing a finished written report. The examples are based on the kinds of preliminary notes made during practical, field course and research project exercises by our own students at various stages of their undergraduate training. They come from four different fields of study, but their common feature is that they provide scope for open-ended investigation and hypothesis testing. Of course, the fact that we happen to have selected these particular examples to illustrate the process is irrelevant to the aim of the book, as the range of other examples running through it amply demonstrates. What emerges from the examples applies with equal weight in all branches of biology from molecular genetics and cell biology to psychology and comparative anatomy.

Example 1

Material: Samples of leaves collected in the field from early successional (dandelion, *Taraxacum officinale*; plantain, *Plantago lanceolata*; poppy, *Papaver rhoeas*), mid successional (clover, *Trifolium repens*; ox-eye daisy, *Leucanthemum vulgare*) and late successional (dogrose, *Rosa canina*; ragwort, *Senecio jacobaea*; blackthorn, *Prunus spinosa*; goldenrod, *Solidago canadensis*) plant species; graph paper; binocular microscope.

Notes: Collected leaf samples show a lot of variation in damage. Some leaves have several semi-circles eaten in from the edges; some have numerous holes through the tissue, many brown round the edges; others show extensive damage with most of the leaf missing in many cases and damage to the stems and twigs they are on. There is

considerable size variation in the leaves. Look at the size range in undamaged leaves and divide into three size classes. Measure size by drawing round whole leaves on graph paper and counting the squares within the boundary. Count number of leaves with different kinds of damage in each size class: 'small' leaves – 41 with small hole damage, 4 with marginal damage, 17 with severe damage; 'medium-sized' leaves – 12 with small holes, 29 with marginal damage, 34 with severe damage; 'large' leaves – 2 with small holes, 22 with marginal damage, 40 with severe damage. One obvious possibility is that size reflects height off the ground and the effects of different kinds of herbivore, perhaps mainly slugs and snails on the smaller leaves, caterpillars and other insects on the medium-sized (shrubs, bushes?) leaves and maybe cattle or deer on the largest. It also looks as if the smaller (low-growing?) leaves have much more damage than the larger ones (84 per cent showing some damage, compared with 47 and 23 per cent in the other two categories). Maybe this is just because bigger plants have more leaves, so more escape damage. While examining the samples, notice several other things. Some leaves have tough 'skins', often with hairs on the surface; these are mainly from the medium-sized category (per cent with hairs visible to naked eye: small, 0; medium, 21; large, 9). Some of the bigger leaves smell strongly or have sticky or latexy sap when squeezed, but the last is also the case with what look like dandelion leaves. The stems of some of the medium-sized and bigger ones have thorns or sticky hairs. There's also more colour variation in these two categories, some leaves being reddish rather than green. It looks like the tougher, more strongly smelling samples generally have less damage than the others.

Material: Stained blood smears, vials of preserved ectoparasites, gut nematodes and faecal samples from live-trapped bank voles (*Clethrionomys glareolus*), microscope with eyepiece graticule, clean microscope slides and coverslips, pipettes.

Example 2

Notes: Looking at blood smears under a microscope, notice range of red and white blood cells. Some red cells in some of the smears have small stained bodies in them. These turn out to be a stage in the life cycle of a protozoan parasite, *Babesia microti*, that infects voles. Some slides seem to have much higher densities of red cells than others. The number of fleas and ticks in each vial varies a lot. Several voles didn't seem to have any, while some had a large number. Divide

samples by age and sex, and do some counts of infected red cells; scan a roughly standard-width field along the graticule scale for each vole and count number of infected cells (adult males: 21, 45, 3, 0, 64; adult females: 16, 1, 13, 0, 0; juvenile males: 0, 5, 34, 0, 0; juvenile females: 0, 0, 0, 16, 0). Count ticks recovered from same groups (adult males: 0, 8, 3, 0, 7; adult females: 0, 1, 0, 0, 2; juvenile males: 2, 2, 0, 5, 3; juvenile females: 0, 0, 0, 2, 0). Smear some faecal samples onto slides and inspect under microscope. Lots of fragments of plant material and detritus. Some samples have clear oval objects which, on asking, turn out to be nematode eggs. Sometimes there are very large numbers of eggs, sometimes none. Too difficult to count all the eggs in each sample, so designate a rank score from 0 (no eggs) to 5 (more than a hundred eggs). Scores for adult males: 3, 5, 5, 0, 3; for adult females: 2, 5, 0, 2, 2; for juvenile males: 2, 2, 4, 2, 1; for juvenile females: 0, 1, 0, 2, 1. On looking at the tubes of preserved worms from the same animals, notice that those with high egg scores do not always have more worms, but some have a greater range of worm sizes. Measure small samples of worms against scale on graticule (ranges for five males: 12–25 units, 9–18 units, 13–28 units, 17–22 units, 11–14 units; ranges for five females: 14–23, 13–19, 12–20, 13–30, 15–29 units).

Example 3

Material: Vials of water containing suspensions of soil-dwelling nematodes from three sites differing in heavy metal and organophosphate pollution, microscope with eyepiece graticule, clean microscope slides and coverslips, pipettes, diagnostic key to common species morphologies.

Notes: Pipette 2×0.2-ml droplets of each sample onto a clean slide and examine under microscope. Count adult worms of identifiable species on each slide. Sample 1 (heavy metal polluted site) six apparent species, call A–F for the moment: numbers – A 27, B 5, C 17, D 3, E 32, F 2; Sample 2 (unpolluted) – B 43, C 4, D 15, F 18, plus four different species G 20, H 31, I 4, J 12; Sample 3 (organophosphate polluted site) – A 5, C 48, E 19, H 11. Juvenile worms also present

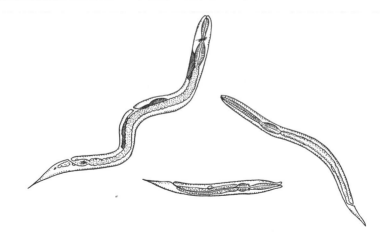

but not readily identifiable to adult species. Nevertheless, numbers in each sample are 23 (Sample 1), 31 (Sample 2), 0 (Sample 3). Can also detect adult females with eggs. Number of females with eggs of each species in samples: Sample 1 – A 6, B 0, C 4, D 0, E 5, F 0; Sample 2 – B 13, C 0, D 2, F 6, G 5, H 11, I 0, J 1; Sample 3 – A 0, C 12, E 7, H 5. Samples contain quite a lot of detritus, some of which can be identified as decomposing nematode material but not related to species. Take ten arbitrarily chosen fields per slide and see how many contain at least one piece of decomposing material: Sample 1 – five, Sample 2 – three, Sample 3 – seven. Repeat for a further two droplets per sample. Number of adults in second set: Sample 1 – A 15, C 21, D 9, F 4; Sample 2 – B 31, C 6, D 21, F 11, G 16, H 24, J 3; Sample 3 – A 8, C 35, D 3, E 14, H 6. Juveniles per sample – 17 (Sample 1), 19 (Sample 2), 5 (Sample 3). Females with eggs: Sample 1 – A 3, C 4, D 0, F 0; Sample 2 – B10, C 0, D 8, F 4, G 4, H 7, J 0; Sample 3 – A 0, C 11, D 0, E 5, H 0. Number of fields per slide with decomposing material: Sample 1 – three, Sample 2 – three, Sample 3 – nine.

Material: Stock cage of virgin female and stock cage of virgin male field crickets (*Gryllus bimaculatus*), two or three 30 × 30-cm glass/Perspex arenas with silver sand substrate, dish of water-soaked cottonwool and rodent pellets, empty egg boxes, assorted colours of enamel paint, fine paintbrush, paint thinners, rule, bench lamps.

Example 4

Notes: Females are distinguished from males by possession of long, thin ovipositor at the back. Put four males into an arena. After rushing about, males move more slowly around the arena. When they meet, various kinds of interaction occur. Interactions involve variety of behaviours: loud chirping, tapping each other with antennae, wrestling and biting. Interactions tend to start with chirping and antenna tapping, and only later progress to fighting. Count number of encounters that result in fighting (15 out of 21). Put in three more males so seven in total and count fights again (8 out of 25). Take out males and choose another five. Take various measurements from each male (length from jaws to tip of abdomen, width of thorax, weight) and mark each one with a small, different-coloured dot of paint. Introduce individually marked males into arena. Count number of fights initiated by each male per encounter (red, thorax width 6.5 mm – 4/10; blue, width 7.5 mm – 8/11; yellow, width 7.0 mm – 8/10; silver, 6.0 mm – 3/12; green, 6.5 mm – 3/9). Continue observations and count number of encounters won by each male (win decided if opponent backs off) (red, 3 wins/6 encounters; blue, 4/5; yellow, 7/7; silver, 1/4; green, 4/8).

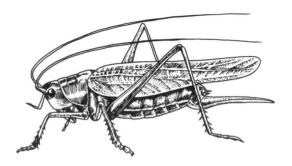

Introduce three sections of egg box to arena and leave males for 10 minutes. After 10 minutes, some males (blue, yellow, red) hiding under egg box shelters or sitting close to them. Males in or near boxes chirping frequently. Count number of encounters resulting in fight (12 out of 16). Count fights/encounter at different distances from burrows for each of the four males in turn – within 5 cm: 4 attempts in 6 encounters for yellow, 2/2 for red, 0/3 for silver, 4/5 for blue; between 5 and 10 cm: 5/11 (yellow), 2/3 (red), 1/5 (silver), 2/2 (blue); 10–15 cm: 3/7 (yellow), 4/8 (red), 0/2 (silver), 3/6 (blue); 15–20 cm: 2/9 (yellow), 1/6 (red), 1/5 (silver), 2/4 (blue); 20–25 cm: 1/8 (yellow), 1/12 (red), 0/2 (silver), 2/8 (blue). Introduce two females. Females move around the arena but end up mainly around the box sections. Show interest in males and occasionally mount. Other males some-times interfere when female mounting particular male. Number of approaches to, and number of mounts with, different males: red, two approaches, no mounts; blue, two approaches, two mounts; yellow, three approaches, two mounts; silver, no approaches; green, one approach, one mount.

2.2   Exploratory analysis

Observational notes are, in most cases, an essential first step in attempting to investigate material for the first time. However, as the examples amply demonstrate, they are a tedious read and, as they stand, do not make it easy to formulate hypotheses and design more informative investigations. What we need is some way of distilling the useful information so that points of interest become more apparent. If we have some numbers to play with – and this underlines the usefulness of making a few measurements at the outset – we can perform some exploratory analyses.

Exploratory analysis may involve drawing some simple diagrams or plotting a few numbers on a scattergram or it might involve calculating some *summary* or *descriptive statistics*. We shall look at both approaches shortly using information from the various sets of observational notes. These sorts of analyses almost always repay the small effort demanded but, like much basic good practice in any field, they are often the first casualty of impatience or prejudgement of what is interesting or to be expected. It is always difficult to discern pattern simply by 'eyeballing' raw numbers and the more numbers we have the more difficult it becomes. A simple visual representation like a scattergram or a bar chart, however, can turn the obscure into the obvious.

2.2.1   Visual exploratory analysis

There are several instances in the examples of observational notes where similar measurements were made from different kinds of material or under different conditions. For instance, infected blood cell counts were taken from adult and juvenile voles of both sexes, while fighting in male crickets was observed at different distances from artificial burrows. In both cases there seem to be some differences in the numbers recorded from different kinds of material (age and sex of host) or under different conditions (distance from a burrow). What do the differences suggest?

Eyeballing the blood cell data suggests some differences both between males and females and between adults and juveniles. A simple way to visualize this might be to total up the number of infected cells scored for each category of animal and present them in a bar chart (Fig. 2.1). If we do this, it looks as though males generally have higher infection levels than females, regardless of age, and that adults have more infection than juveniles, regardless of sex. From this, we might be tempted to suggest that adult males are particularly prone to infection compared with other classes of individual. As we shall see later, however, we might want to be cautious in our speculation.

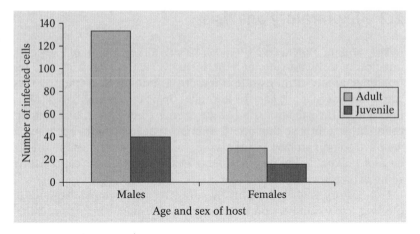

Figure 2.1 The total number of infected red blood cells in voles of different age and sex (*see* Example 2, Notes).

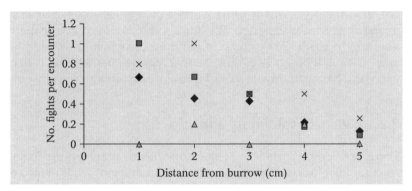

Figure 2.2 The number of fights per encounter between crickets at different distances from the nearest burrow; symbols represent individuals (*see* Example 4, Notes).

In the cricket example, we can see that the number of fights per encounter seems to vary with distance away from an artificial burrow. An obvious way to see how is to plot a scattergram of number of fights per encounter against distance. The result (Fig. 2.2) suggests that, while there is a fair spread of values at each distance, there is a tendency for more encounters to result in a fight when crickets are close to a burrow. It is important to bear in mind that, as each cricket was observed in turn, results for the different animals are independent of each other and the trend is not a trivial outcome of a fight scored for one cricket also counting as a fight for his opponent. As we shall see later, this and other kinds of nonindependence can be troublesome in drawing inferences from data.

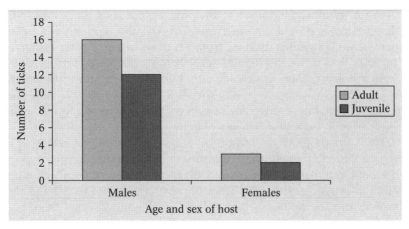

Figure 2.3   The number of ticks recovered from voles of different age and sex (*see* Example 2, Notes).

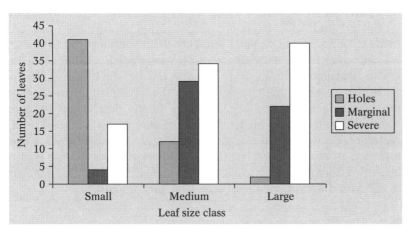

Figure 2.4   The number of small, medium and large leaves showing different kinds of damage by herbivores (*see* Example 1, Notes).

These are just two examples. We could do similar things with various other measurements in the observational notes. Figure 2.3, for instance, suggests that the number of *Babesia*-infected cells in different classes of host might have something to do with tick burden since both show the same broad association with class (*see* Fig. 2.1). Figure 2.4 hints at an association between the size of a leaf and the kind of damage it sustains. The Notes provide scope for other exploratory plots; try some for yourself.

Casting data in the form of figures like this is helpful not just because visual images are generally easier for most people to assimilate than raw numbers, but because they can expose subtleties in the data that are less apparent in numerical form. The plot of number of fights per encounter against distance from a burrow in Fig. 2.2, for instance, suggests that while the likelihood of fighting decreases further away from a burrow, there is considerable individual variation (the different symbols) within the trend. We shall see later that this variation leads to some interesting insights into the role of burrows in the behaviour of both male and female crickets.

The sorts of plots we have used so far are helpful in seeing at a glance whether something interesting might be going on. However, the data could be presented in a different way to make the figures more informative. It is clear from the scattergram in Fig. 2.2 and the raw data in the notes relating to Figs 2.1, 2.3 and 2.4 that there is a lot of variation in the numbers recorded in each case. Adult male voles, for example, did not all have high *Babesia* burdens; indeed one male didn't have any infected cells at all in the sample examined. This variability has at least two important consequences as far as exploratory plots are concerned. First it suggests that simply plotting totals in Figs 2.1 and 2.3 is likely to be misleading, because the totals are made up from a wide range of numbers. A large total could be due to a single large result, with all the rest actually being smaller than the results contributing to the other, lesser, totals, in which case our interest in the apparent differences between the bars in the figures might diminish somewhat. Second, variability in the data might obscure some potentially interesting tendencies in scattergrams. What we need, therefore, is a way of summarizing data so that: (a) the interesting features are still made clear, but (b) the all-important variability is also presented, though in a way that clarifies rather than obscures patterns in the data. In short, we need some *summary statistics*.

2.2.2 Summary statistics

The usual way of summarizing a set of data so as to achieve (a) and (b) above is to calculate a *mean* (average) or *median* value and then to provide as a measure of the variability the associated *standard error* (for a mean) or *confidence limits* (for a median).

Means and standard errors
The mean (often represented by $\bar{x}$ ('*x*-bar')) is simply the sum of all the individual values in the data set divided by the number of values (usually referred to as n, the sample size). Formally, the mean is expressed as:

$$\bar{x} = (1/n) \sum_{i=1}^{i=n} x_i$$

The expression $\sum_{i=1}^{i=n} x_i$ indicates that the first ($i = 1$) to the nth ($i = n$) data values (x) are summed (Σ is the summation sign) and can be expressed more simply as Σx. This summed value is then multiplied by $1/n$ (equivalent to dividing by the sample size n).

Since the mean is calculated from a number of values, we need to know how much confidence we can have in it. By 'confidence' we mean the reliability with which we could take any such set of values from the material and still end up with the same mean. A statistician would phrase this in terms of our sample mean ($\bar{x}$, the one we have calculated) reflecting the true mean (usually denoted μ) of the population from which the data values were taken. Suppose, for instance, we measured the body lengths of ten locusts caught in each of two different geographical areas and obtained the following results:

Length (cm)	
Area 1	Area 2
6.3	6.0
7.1	6.4
6.2	6.3
6.5	6.0
7.0	5.9
6.7	6.5
6.5	6.1
7.0	6.2
6.8	6.2
7.1	6.4
$\bar{x}$ 6.7	6.2

Then suppose that we had obtained the following instead:

Length (cm)	
Area 1	Area 2
8.3	8.0
5.1	5.4
7.2	5.3
5.5	6.5
8.0	8.4
5.7	5.5
5.5	7.1
8.0	5.2
8.8	5.2
5.1	5.4
$\bar{x}$ 6.7	6.2

In both cases, the mean body lengths from each area are the same and we might want to infer that there is some difference in body size between areas. In the first case, the range of values from which each mean is derived is fairly narrow but different between areas. We might thus be reasonably happy with our inference. In the second case, however, the values vary widely within areas and there is considerable overlap between them. Now we might want to be more cautious about accepting the means as representative of the different areas. We can see that this is the case from the columns of numbers but we need some way of summarizing it without having to present raw numbers all the time. We can do this in several ways. The most usual is to calculate the standard error, a quantitative estimate of the confidence that can be placed in the mean and which can be presented with it as a single number. The calculation is straightforward and is shown in Box 2.1.

Box 2.1	Standard deviations and standard errors

The *standard deviation* (usually abbreviated to s.d. or SD) measures the spread of actual data values around the mean, on the assumption that these data follow the *normal distribution* (*see* p. 48); it is a measure of the confidence you have that any particular data value will fall within a particular range (the mean + 1 s.d. and the mean − 1 s.d., hence the mean ± s.d.). The standard error of the mean, usually just called the *standard error* (abbreviated to s.e. or SE) measures the spread of multiple *sample means* around the true population mean. Normally you will only be taking a single sample, and hence the SE is an expression of the confidence you have that your sample mean falls within a particular range (mean ± s.e.) of the true population mean. Since sample means are almost always normally distributed, it is always OK to cite a standard error with your sample mean. The calculations are as follows:

The standard deviation:

1. Calculate the sum of all the data values in your group (Σx).

2. Square the individual data values and sum them, giving (Σx^2).

3. Calculate $\Sigma x^2 - (\Sigma x)^2/n$ (remember that n is the sample size, the number of values in your set of data). This is actually a quick way of calculating the sum of squared deviations from the mean: $\Sigma(x - \bar{x})^2$. The deviations are squared so that positive and negative values do not simply cancel each other out.

4. Dividing by $n - 1$ gives the *variance* of the sample, an important intermediate quantity in many statistical tests.

5. Taking the square root of the variance gives the *standard deviation*.

The standard error:

1. Follow steps 1–4 of the calculations for the s.d. to obtain the variance of the sample.

2. Divide by n, the sample size (despite the fact that you have just divided by $n-1$ to get the variance!).

3. Taking the square root of the result will then give you the *standard error*.

Most scientific calculators will give you the mean of a set of numbers and most will also give you the *standard deviation* (Box 2.1), usually represented as σ or s. If your calculator has both σ and σ_{n-1} buttons, it is the σ_{n-1} one that you want. The standard deviation will become important later, but for the moment we can simply use it to obtain the standard error. All we need to do is call up the standard deviation, square it, divide it by n and take the square root.

Whichever way you calculate the standard error (by hand or by calculator), it should be presented with the mean as follows:

$$\bar{x} \pm \text{s.e.}$$

The $\pm$ sign indicates that the standard error extends to its value on either side of the mean. The bigger the standard error, therefore, the more chance there is that the true mean is actually greater or smaller than the mean we've calculated. We can see how this works with our locust data. Let's look at the two sets of values for Area 1, first calculating the x^2 value of the first example:

Locust	Body length (x)	x^2
1	6.3	39.69
2	7.1	50.41
3	6.2	38.44
4	6.5	42.25
5	7.0	49.00
6	6.7	44.89
7	6.5	42.25
8	7.0	49.00
9	6.8	46.24
10	7.1	50.41
$n = 10$	$\Sigma x = 67.2$	$\Sigma x^2 = 452.58$

The steps are then:

1. $\Sigma x = 67.2$
2. $\Sigma x^2 = 452.58$
3. $\Sigma x^2 - (\Sigma x)^2/n = 452.58 - (67.20)^2/10 = 452.58 - 451.58 = 1$
4. Divide by $n - 1 = 1/9 = 0.11$
5. Divide by $n = 0.11/10 = 0.01$
6. Take the square root $= \sqrt{0.01} = 0.11$

Thus the mean length of locusts in the first example for Area 1 is:

6.72 ± 0.11 cm

If we repeat the exercise for the second example, however:

1. $\Sigma x = 67.2$
2. $\Sigma x^2 = 471.18$
3. $\Sigma x^2 - (\Sigma x)^2/n = 471.18 - 451.58 = 19.6$
4. Divide by $n - 1 = 19.6/9 = 2.2$
5. Divide by $n = 2.2/10 = 0.22$
6. Square root $= \sqrt{0.22} = 0.47$

The mean is now expressed as:

6.72 ± 0.47 cm

We could leave the mean and standard error expressed numerically like this, or we could present them visually in a bar chart. If we opt for the bar chart, then the mean can be plotted as a bar and the standard error as a line through the top centre of the bar extending the appropriate distance (the value of the standard error) above and below the mean. Figure 2.5a, b shows such a plot for the two sets of example locust data.

Medians and confidence limits

An alternative summary statistic we could have used is the median. There are good statistical reasons, to do with the distribution of data values (*see* Chapter 3), why we may need to be cautious about using means. The use of mean values makes important assumptions about the distribution of the data from which they are calculated that may not hold in many cases. Using medians avoids these assumptions. Later, we shall see that statistical tests of significance can also avoid

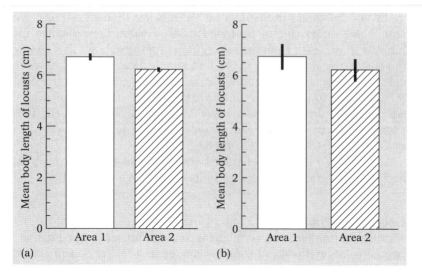

Figure 2.5

them. Finding the median is simple. All we do is look for the central value in our data. Thus, if our data comprised the following values:

5   7   11   21   8   12   14

we first rank them in order of increasing size:

5   7   8   11   12   14   21

and take the value that ends up in the middle, in this case 11. If we have an even number of values so there isn't a single central value, we take the halfway point between the two central values, thus in the following:

2   6   8   20   23   38   40   85

the median is 21.5 (halfway between 20 and 23). Note that the median may yield a value close to or very different from the mean. In the first sample, the mean is 11.1 and thus similar to the median. In the second sample, however, it is much greater at 27.58.

Again, we want some way of indicating how much confidence to place in the median. By far the simplest way is to find the confidence limits to the median using a standard table, part of which is shown in Appendix I. All we need to do is rank order our data values as before, count the number of values in the sample (n), then use n to read off a value r from the table. Normally, we would be interested in the r value appropriate to confidence limits of approximately 95 per cent ('approximately' because, of course, the limits always have to be two of the values in the data set – if n is less than six, 95 per cent confidence limits cannot be found). This r value then dictates the number of values in from the two extremes of the data set that denote the confidence limits. Thus in our first sample data set, there are seven values. Reference to Appendix I shows that for $n = 7$, $r = 1$; the confidence limits to the median of 11 are therefore 5 and 21. However, if we had a sample of nine values (say 7, 11, 15, 22, 46, 67, 71, 82, 100) r for approximately 95 per cent confidence limits is 2, so for a median of 46, the limits would be 11 and 82.

As with the mean and standard error, we can represent medians and their associated confidence limits visually in a bar chart.

Frequency distributions

Means and standard errors, medians and confidence limits, then, are two conventional ways of summarizing central tendency and spread of values within data. As we have seen, they are particularly useful in making quick 'eyeball' comparisons between two or more data sets. Such 'eyeball' comparisons, however, are only one reason why visual summaries of data sets can be worth plotting. Another is to allow the features of a single set of data to be explored fully, perhaps to examine the distribution of values and decide on an appropriate method of confirmatory analysis (*see later*). The usual way of doing this is to plot a *frequency distribution* of the values (the number of times each occurs in the data set), either as a bar chart, like the means in Fig. 2.6, or as a histogram. Bar charts, in which the bars in the figure are separated by a small gap along the horizontal (x) axis, are used to plot

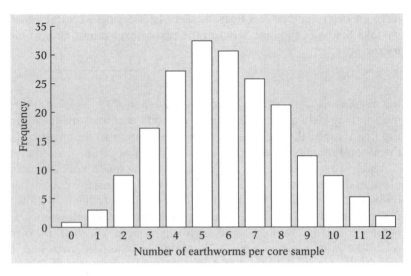

Figure 2.6 A bar-chart frequency distribution of different numbers of earthworms recovered from core samples of soil (modified from Dytham, 1999).

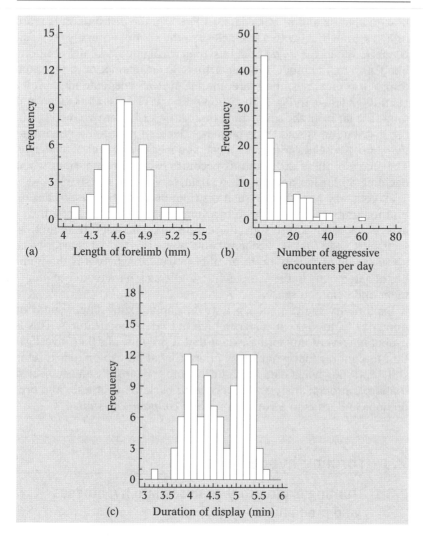

Figure 2.7 Histogram frequency distributions of: (a) the length of the right forelimb of field cricket (*Gryllus bimaculatus*) nymphs, (b) the number of aggressive encounters per day between male mice (*Mus musculus*) in an enclosure and (c) the duration of the aggressive displays of male Siamese fighting fish (*Betta splendens*) in pairwise encounters.

the distribution of discrete values where there are sufficiently few of these to make such a plot feasible. Histograms, in which the bars are contiguous, are used to plot the distribution of classes of values (e.g. 1–10, 11–20, 21–30, etc.), usually where there is a wide and continuous spread of individual values. Figure 2.7 shows a range of frequency distribution histograms. It is obvious almost at a glance why we urged caution in using the mean as a general measure of central tendency and why frequency distributions can be a crucial first step in deciding how to analyse and present data. In Fig. 2.7a the distribution is more or less symmetrical with a peak close to the centre. Formally the peak is referred to as the *mode* and, in symmetrical, unimodal distributions like Fig. 2.7a, the mode and the arithmetic mean amount to the same thing. In these cases, therefore, the mean is an adequate measure of central tendency. In Fig. 2.7b, c, however, it is clear that the mean and mode are far from the same thing (in fact Fig. 2.7c has two modes and the distribution is said to be bimodal). Here, it makes little quantitative sense to use the arithmetic mean as a measure of central tendency. The shape of these distributions becomes extremely important when deciding on further analysis of the data, as we shall see shortly.

Of course, when plotting such distributions, one has first to decide on the number of categories into which to cast the data along the *x*-axis. Perhaps not surprisingly, there is no hard and fast rule about this. Dytham (1999), in his excellent book, advocates 12–20 categories as a rough rule of thumb, or, as an alternative, $\sqrt{n}$, where n is the number of data points in the sample. But, in the end, it's a matter of judgement and common sense.

Exploratory analysis, then, is a way of summarizing data, visually or numerically, to make it easier to pick out interesting features. This is useful, however, only to the extent that it promotes further investigation to confirm that what looks interesting at an exploratory level is still interesting when data are collected more rigorously and subjected to more thorough analysis. This brings us back to hypotheses and predictions and leads to a consideration of *confirmatory statistics*.

2.3   Forming hypotheses

2.3.1   Turning exploratory analyses into hypotheses and predictions

Exploratory analyses are generally open-ended in that they are not guided by preconceived ideas about what might be going on. However, they are the first important step on the way to formulating hypotheses which do then guide investigation. As we have seen already, hypotheses can be very general or they can be specific. Both kinds can be generated from our observational notes.

Example 1

Plants and herbivores

The observations of leaf damage in the samples of plant species suggest that several factors are influencing the type and amount of damage. A possibility emerging from the Notes is that damage decreases with height off the ground and thus vulnerability to slugs. This can be framed as a readily testable hypothesis:

Hypothesis 1A *The type and extent of leaf damage reflects availability to slugs.*

from which some predictions for testing might be:

Prediction 1A(i) *Taller plant species will have less leaf damage by slugs than shorter species.*

Prediction 1A(ii) *Leaf damage will decrease the further up a plant that samples are taken.*

Of course, this is a very broad hypothesis and many features of a leaf affecting its likelihood of predation are going to change along with its height off the ground. Some of these are suggested by other observations. For example:

Observation Larger leaves are often tougher or smell strongly.

Hypothesis 1B *The decrease in damage among larger leaves is due to their reduced palatability.*

Prediction 1B *For any given size of leaf, damage will decrease the tougher the cuticle or the stronger the odour on crushing.*

Observation Larger leaves are sometimes associated with thorns or sticky hairs on the stems.

Hypothesis 1C *Reduced damage among larger leaves is due to grazing deterrents on the stems.*

Prediction 1C(i) *The incidence of severe damage (suggestive of large herbivores) will be lower on thorny species.*

Prediction 1C(ii) *The incidence of less severe damage (suggestive of invertebrate herbivores) will be lower on species with sticky, hairy stems.*

The notes on the samples of material and parasites taken from voles suggest a number of interesting possibilities, some to do with the age and sex of the voles, others to do with relationships between the different parasites.

Example 2

Hosts and parasites

Observation The number of *Babesia*-infected red cells and faecal egg scores appeared to be higher in male voles than in females, and higher in adults than in juveniles.

Hypothesis 2A *Parasite burdens are affected by differences in the levels of reproductive hormones between age and sex classes.*

Prediction 2A *Parasite burdens will increase with host testosterone levels.*

Hypothesis 2B   *Parasite burdens are affected by differences in aggressive behaviour and stress between age and sex classes.*

Prediction 2B(i)   *Parasite burdens will be greater among dominant territorial males.*

Prediction 2B(ii)   *Parasite burdens will increase in any individual with the amount of aggressive behaviour shown.*

Prediction 2B(iii)   *Parasite burdens will increase with host corticosterone (stress hormone) levels.*

Observation   Sex and age differences in *Babesia* levels appear to be associated with the number of ticks on the host.

Hypothesis 2C   *The intensity of infection with* Babesia *depends on the degree of exposure to infected ticks.*

Prediction 2C   Babesia *burden will increase with the number of infected ticks establishing on the host.*

Hypothesis 2D   *The intensity of infection with* Babesia *depends on the degree of resistance to tick infection.*

Prediction 2D   *The intensity of infection with* Babesia *will decrease with the host's ability to mount an antibody response to ticks.*

| **Example 3** |
| **Nematodes and pollutants** |

Observations on the samples of soil-dwelling nematodes suggest a number of things vary with the pollution status of the site of origin – species diversity and fecundity among them.

Observation   Fewer species were identified in the samples from the two polluted sites compared with the unpolluted site.

Hypothesis 3A   *Pollution reduces species diversity.*

Prediction 3A   *The addition of pollutants to identical multi-species nematode cultures will result in a reduction in the number of species supported over time.*

Hypothesis 3A is another very broad hypothesis and could give rise to several more specific hypotheses each generating their own predictions. For example:

Hypothesis 3B   *Pollutants are toxic to those species missing from polluted sites.*

Prediction 3B   *Species present at unpolluted sites but missing from polluted sites will show greater mortality when exposed to pollutants.*

Hypothesis 3C *Pollutants affect resource availability for certain groups of nematodes.*

Prediction 3C *Species missing from polluted sites will tend to come from certain trophic or microhabitat groups.*

Observation Some species are present only in polluted sites.

Hypothesis 3D(i) *Such species benefit from relaxed interspecific competition in polluted sites.*

Prediction 3D(i) *Increasing the number of species in the culture in an otherwise constant and pollutant-free environment will tend to result in the loss of such species from the community.*

Hypothesis 3D(ii) *Pollutants create niche opportunities not available in unpolluted sites.*

Prediction 3D(ii) *For any given number of species in the culture in an otherwise constant environment, such species will do better when pollutant is added compared with an unpolluted control.*

Observation Fewer juvenile stages were recorded from the organophosphate polluted site than from the other two sites, but there was no consistent difference in the number of females with eggs.

Hypothesis 3E(i) *Organophosphate pollution affects recruitment to nematode populations through reduced egg viability.*

Prediction 3E(i) *Females reared in organophosphate-treated, single-species culture will show reduced hatching success per egg compared with those reared in heavy metal or untreated control cultures.*

Hypothesis 3E(ii) *Organophosphate pollution affects recruitment to nematode populations through increased juvenile mortality.*

Prediction 3E(ii) *Females reared in organophosphate-treated, single-species culture will show comparable hatching success per egg but reduced survival of resultant juvenile stages than those reared in heavy metal or untreated control cultures.*

Male field crickets seemed to be aggressive to one another when put together in an arena. Whether or not an encounter resulted in fighting varied with the number of crickets, and individuals differed in their tendency to initiate and win fights. The apparent effects of providing egg box shelters and introducing females suggest that interactions between males are concerned ultimately with gaining access to females.

Example 4

Crickets

Observation   The number of encounters leading to a fight was lower when more crickets were present.

Hypothesis 4A   *The cost of fighting on encounter increases with population size and the chance of encountering another male.*

Prediction 4A   *The probability of an encounter's resulting in a fight will decrease with increasing numbers of males and in the same number of males maintained at a higher density.*

Observation   Larger males initiated more fights per encounter and won in a greater proportion of encounters.

Hypothesis 4B   *Large size confers an advantage in fights between males.*

Prediction 4B   *Males will be less likely to initiate a fight when their opponent is larger.*

Observation   Interactions tended to escalate from chirping and antenna-tapping to overt fighting.

Hypothesis 4C   *The escalating sequence reflects information-gathering regarding the size of the opponent and the likelihood of winning.*

Prediction 4C　*Encounters will progress further when opponents are more similar in size and it is more difficult to judge which will win.*

Observation　Larger males ended up in or near egg box shelters and females tended to spend more time with these males.

Hypothesis 4D(i)　*Females prefer to mate with males in shelters for protection from predators.*

Prediction 4D(i)　*Giving a male a shelter will increase the attention paid to him by females and his chances of copulating.*

Hypothesis 4D(ii)　*Females prefer large males.*

Prediction 4D(ii)　*Given a choice of males, all with or without shelters, females will spend more time and be more likely to copulate with larger males.*

2.3.2　Null hypotheses

In the examples above, we have phrased predictions in terms of the outcomes they lead us to expect. Prediction 1B, for example, leads us to expect that the amount of damage sustained by a leaf will decrease the tougher its cuticle or the more volatiles it contains. We can test this prediction by carrying out a suitable investigation and associated confirmatory analysis. Formally, however, we do not test predictions in this form. Rather, we test them in a null form that is expressed as a hypothesis *against* the prediction. This is known as a *null hypothesis* and is often expressed in shorthand as H_0. Predictions are tested in the form of a null hypothesis because science proceeds conservatively, always assuming that something interesting is *not* happening unless convincing evidence suggests, for the moment, that it might be. In the case of Prediction 1B, therefore, the null hypothesis would be that tougher cuticles or more volatiles would make no difference to the amount of damage sustained by a leaf. We shall see later what burden of proof is necessary to enable us to reject the null hypothesis in any particular case.

There is a second, and from a practical point of view more crucial, point to make about the predictions. Skimming down them gives the impression of specificity and diversity; each prediction is tailored to particular organisms and circumstances, and those from one example seem to have little to do with those from others. At the trivial level of detail, this is obviously true. However, in terms of the kinds of question they reflect, predictions from the different examples in fact have a great deal in common. Before we can proceed with the problem of testing hypotheses and choosing confirmatory analyses, we need to be aware of what these common features are.

2.3.3   Differences and trends

Although we derived some 23 different predictions from our notes, and could have derived many more, all fall (and any others would have fallen) into one of two classes. Regardless of whether they are concerned with nematodes or crickets or with surviving pollutants or fighting rivals, they either predict some kind of *difference* or they predict some kind of *trend*. Recognizing this distinction is vitally important because it determines the kind of confirmatory test we shall be looking to perform and therefore the design of our experiments. Surprisingly, however, it proves a stubborn problem for many students throughout their course, with the result that confirmatory analyses often fall at the first fence. Let's look at the distinction more closely.

A *difference* prediction is concerned with some kind of difference between two or more groups of measurements. The groups could be based on any characteristics that can be used to make a clear-cut distinction; obvious examples could be sex (e.g. a difference in body size between males (Group 1) and females (Group 2)), functional anatomy (e.g. a difference in enzyme activity between xylem (Group 1), phloem (Group 2) and parenchyme (Group 3) cells in the stem of a flowering plant), or experimental treatment (e.g. a difference in the number of chromosomal abnormalities following exposure to a mutagen (Group 1) or exposure to a harmless control (Group 2)). Which of the predictions we derived earlier are difference predictions?

Example 1

Plants and herbivores

In the leaf sample study there are two difference predictions:

- Prediction 1C(i) leads us to expect that the incidence of severe damage will be lower on thorny species (Group 1) than on non-thorny species (Group 2).

- Prediction 1C(ii) leads us to expect a reduction in less severe damage on species with sticky, hairy stems (Group 1) compared with species without (Group 2).

Example 2

Hosts and parasites

One prediction from the example of host/parasite relationships involves a difference:

- Prediction 2B(i) suggests a difference in parasite burden between dominant territorial males (Group 1) and other age and sex categories of host (Group 2).

Almost all the predictions arising from the soil-dwelling nematode samples turn out to be difference predictions:

Example 3

Nematodes and pollutants

- Prediction 3A suggests a difference in the number of species between cultures to which pollutant has been added (Group 1) and those that are pollutant-free (Group 2).

- Prediction 3B predicts a difference in sensitivity to pollutants between species absent from polluted sites (Group 1) and those present at such sites (Group 2).

- Prediction 3C suggests a difference between trophic groups (e.g. bacterial feeders (Group 1), fungal feeders (Group 2), plant feeders (Group 3)) in the tendency to be present at polluted sites.

- Prediction 3D(ii) leads us to expect a difference between cultures to which pollutant has been added (Group 1) and pollutant-free controls (Group 2) in the tendency to support nematode species found only in polluted sites in the field.

- Prediction 3E(i) suggests a difference in egg hatching success between female worms exposed to organophosphate pollutant (Group 1) and those not (Group 2).

- Prediction 3E(ii) is similar to the last prediction except that it suggests a difference in larval mortality instead.

Two predictions from the crickets involve differences:

Example 4

Crickets

- Prediction 4B suggests that males will be less likely to intitiate a fight when their opponent is larger than them (Group 1) than when it is smaller (Group 2).

- Prediction 4D(i) predicts that females will pay more attention to males with a shelter (Group 1) than to males without a shelter (Group 2).

Trend predictions are concerned not with differences between hard and fast groupings but with the relationship between two more or less continuously distributed measures. Thus, for example, a relationship might be predicted between the amount of an anthelminthic drug administered to a rat infected with nematodes and the number of worm eggs subsequently counted in the animal's faeces. In this case, we should expect the relationship to be negative with egg counts decreasing the more drug the rat has received. On the other hand, a positive relationship might be predicted between the number of hours

of sunlight received and the standing crop of a particular plant. With trends we can therefore envisage two measures as the axes of a graph. One measure extends along the bottom (x) axis, the other up the vertical (y) axis. Sometimes it doesn't matter which measure goes along the x-axis and which up the y-axis because there is no basis for implying cause and effect and we are interested only in whether there is some kind of association. Thus, we might expect a strong association between the amount of ice cream eaten and the amount of time spent in the sea on a visit to the seaside because both would go up with temperature. Since neither could reasonably be thought of as a cause of the other, it is of no consequence which goes on the x- or y-axis. In the two examples above, however, there are reasonable grounds for supposing cause and effect. While it is plausible for the anthelminthic drug to affect faecal egg counts, it is not plausible for the egg counts to have influenced the amount of drug. Similarly, hours of sunlight could influence a standing crop but not vice versa. In these cases, the drug dose and hours of sunlight measures should go on the x-axis and the egg counts and standing crops on the y-axis. It is important to stress, however, that by doing this we are not asserting that the x-axis measure really *is* a cause of the y-axis measure – as we shall see later, inferring cause and effect from relationships requires extreme caution – merely that if there was a cause and effect relationship it would most likely be that way round. This is also clear in the remainder of our example predictions all of which involve trends.

Example 1

Plants and herbivores

In Prediction 1A(i), leaf damage is expected to decrease as the height of plant species increases. Plant height should thus be the x measure and leaf damage the y measure:

- Prediction 1A(ii) makes a similar prediction except that the expected relationship is *within* plants. Height up the plant is the x measure and leaf damage once again the y measure.

- Prediction 1B suggests a negative relationship between toughness of the cuticle (x measure) and leaf damage (y measure).

Example 2

Hosts and parasites

All except one of the predictions from the vole parasites example are trend predictions:

- Prediction 2A predicts an increase in parasite burdens (y measure) with increasing testosterone levels (x measure).

- Predictions 2B(ii) and (iii) predict similar increases in parasite burdens (y measure) but this time as a function of increasing aggression and corticosterone levels (x measures), respectively.

- Prediction 2C suggests an increase in *Babesia* levels (*y* measure) with the number of ticks recovered from the host (*x* measure).

- Prediction 2D suggests a reduction in *Babesia* levels (*y* measure) with host immune responsiveness (*x* measure).

Only one of the predictions arising from the nematode example suggests a trend:

Example 3

Nematodes and pollutants

- Prediction 3D(i) predicts a loss of species found only at polluted sites (*y* measure) as the number of species in a culture increases (*x* measure).

Three predictions from the crickets involve trends:

Example 4

Crickets

- Prediction 4A first of all predicts that the number of encounters ending up in a fight (*y*) will increase with the number of crickets (*x*), then predicts a similar increase when the same number of crickets are maintained at higher densities (density = *x*):
- Prediction 4C involves a predicted trend in the tendency to escalate an interaction (*y*) with decreasing difference in size between opponents (*x*).
- Prediction 4D(ii) suggests that the time females spend with a male (*y*) and their tendency to copulate (*y*) will increase with male size (*x*).

There is thus a clear distinction between difference and trend predictions. Of course, it is possible to recast some trend predictions as difference predictions (for instance, a continuous measure of group size for use in a trend could always be recast in terms of small groups (groups below size *w*) and large groups (groups above size *w*) and thus be used in a difference prediction). What makes the distinction, therefore, is not the data *per se* but the way data are to be collected or classified for analysis. Thus, while measures such as group size or time intuitively suggest trends, there is nothing to stop their being used in difference predictions. It all depends on what is being asked. This is often a source of serious confusion among students encountering open-ended data-handling for the first time.

2.4 Summary

1. Open-ended observation is a good way to develop the basis for forming hypotheses and predictions about material. It pays to make observations quantitative where possible so that exploratory analyses can highlight points of interest.

2. Exploratory analysis is a useful (often essential) first step in extracting interesting information from observational notes or other sources of exploratory information. It can take a wide variety of forms, such as bar charts, scattergrams, or tables of summary statistics.

3. Exploratory analyses, or raw exploratory information itself, can lead to a number of hypotheses about the material. In turn, each hypothesis can give rise to several predictions that test it. Formally, predictions are tested in the form of null hypotheses.

4. While predictions derived from hypotheses may be diverse and specific in detail to the material of interest, they fall into two clearly distinguishable categories: predictions about *differences* and predictions about *trends*. Which of these categories a prediction belongs to is determined by the way data are to be collected or classified for analysis.

Reference

Dytham, C. (1999) *Choosing and using statistics: a biologist's guide.* Blackwell, Oxford.

CHAPTER 3

Answering questions

What do the results say?

In the last chapter, we looked at the way hypotheses can be derived from exploratory information. We turn now to the problem of how to test our hypotheses. As we have seen, we begin by making predictions about what should be the case if our hypotheses are true. These predictions then dictate the experiments or observations that are required. However, this may not be as straightforward as it sounds; decisions have to be made about what is to be measured and how, and how the resulting data are to be analysed. The questions of measurement and analysis are, of course, interdependent. This is obvious both at the level of choosing between difference and trend analyses – there is little point collecting data suitable for a difference analysis if what we're looking for is a trend – and at the level of analyses *within* differences and trends. While at first sight it might seem like putting the cart before the horse, therefore, we shall introduce confirmatory analysis *before* dealing with the collection of data so that the important influence of choice of analysis on data collection can be made clear.

3.1   Confirmatory analysis

3.1.1   The need for a yardstick in confirmatory analysis: statistical significance

Take a look at the scattergram in Fig. 3.1. It shows a relationship between the concentration of a fungicide sprayed on a potato crop and the percentage of leaves sampled subsequently that showed evidence of fungal infection. A plot like this was presented to a class of undergraduates. Students in the class were asked whether they

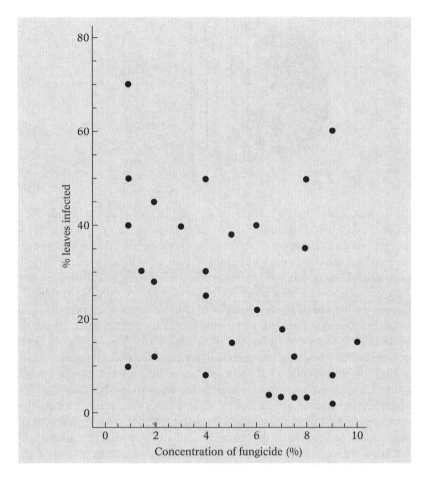

Figure 3.1   A scattergram of the relationship between concentration of a fungicide applied to a potato crop and the percentage of leaves subsequently found to be infected.

thought it suggested any effect of fungicide concentration on infection. The following are some of their replies:

Yes, fungicide concentration obviously has an effect because infection goes down with increasing concentration.

It's hard to say. It looks as though there is some effect, but it's pretty weak. More data are needed.

Fungicide concentration is reducing infection but there must be other things affecting it as well because there's so much scatter.

I don't think you can say anything from this. Yes, there is some downward trend with increasing concentration but several points for high concentrations are higher than some of those for low concentrations. Totally inconclusive.

Yes, there is a clear negative effect.

Clearly there are different, subjective, reactions to the plot. To some it is unequivocal evidence for an effect of fungicide concentration; to others it doesn't suggest much at all. Left to 'eyeball' impressions, therefore, the conclusion that emerged would be highly dependent on who happened to be doing the eyeballing. What is required, quite clearly, is some independent yardstick for deciding whether or not we can conclude anything from the relationship. Since the scenario above could be repeated with any set of data – difference or trend – the need for such a yardstick arises in all cases. One or two idiosyncratic departures notwithstanding (one well-known ornithologist used to advocate the yardsticks 'not obvious', 'obvious' and 'bloody obvious'), the yardstick used conventionally in science is *statistical significance*. There is nothing magical or complicated about statistical significance. It is simply an arbitrary criterion accepted by the international scientific community as the basis for accepting or rejecting the null hypothesis in any given instance and thus deciding whether predictions, and the hypotheses from which they are derived, hold. If the criterion is reached, the difference or trend in question is said to be *significant*; if it is not, the result is *non-significant*. The term 'significant' thus has an important, formal meaning in the context of data analysis and its use in a casual, everyday sense should be avoided in discussions relating to scientific interpretation. How do we decide whether differences or trends are significant? By using the most appropriate of the vast range of significance tests at our disposal. Before we introduce some of these tests, however, we must say a little more about significance itself.

3.2   What is statistical significance?

A *statistic* is a measure, such as a mean or a correlation, derived from samples of data that we have collected. Our expectation is that it relates closely to an equivalent, real value (*parameter*) in the population from which the samples were drawn. Of course it might or it might not. Our sampling technique (*see later*) may have been impeccable and produced a very accurate reflection of the real world. More than likely, however, and for all kinds of forgivable reasons, it will have produced a somewhat biased sample and our calculated statistics will differ from their population parameters. This is what makes statistical inference tricky. If we detect an apparent difference between two sets of data, or an apparent trend in the relationship between them, is the difference or trend real or is it just an artefact of the chance bias in our sampling? Or, to put it another way, is it statistically significant?

The criterion that determines significance is the probability (usually denoted as α) that a difference or trend as extreme as the one observed would have occurred if the null hypothesis – that there is really *no* difference or trend in the population from which the sample came – was true. Confused? An example makes it clear. Let's take one of our earlier predictions, say Prediction 2C. This predicts an increase in *Babesia* burden in voles with increasing degree of tick infestation. Suppose we had tested this by infesting each of ten sets of five parasite-free voles with a different number of *Babesia*-infected ticks, measured the subsequent number of infected blood cells in each vole and found what looked like a convincing positive trend: *Babesia* burden goes up with increasing numbers of ticks. The null hypothesis in this case, of course, is that *Babesia* burden will *not* increase with the number of ticks. What, then, is the probability of obtaining a positive relationship as extreme as the one we got if this null hypothesis is really true and the apparent trend a chance effect? A helpful analogy here might be the probability of obtaining the apparent trend by haphazardly throwing darts at the scattergram. An appropriate significance test will tell us (we shall see how later). By convention in biology, a probability of 5 per cent (= 0.05 when expressed as a proportion) is accepted as the threshold of significance. If the probability of obtaining a relationship as extreme as ours by chance turns out to be 5 per cent or less, we can regard the relationship as significant and reject the null hypothesis. If the probability is greater than 5 per cent we do not reject the null hypothesis and the relationship is regarded as non-significant. If the null hypothesis is not rejected, we effectively assume that our apparent relationship was due to a chance sampling effect. As a matter of interest, the negative trend in Fig. 3.1 is significant at the 5 per cent level, so the optimists have it in this case!

The 5 per cent threshold is, of course, arbitrary and still leaves us with a one-in-twenty chance of rejecting the null hypothesis incorrectly (falsely accepting there is a difference or a trend when there isn't). Under some circumstances, for instance when testing the effectiveness of a drug, a one-in-twenty risk of incorrect rejection might be considered too high. In certain areas of research such as medicine, therefore, the arbitrary threshold of significance is set at 1 per cent ($= 0.01$). In other disciplines it is sometimes relaxed to 10 per cent ($= 0.1$). Although we have talked of threshold probability (p) values ($p < 0.05$, $p < 0.01$, etc.), most computer statistical packages now quote *exact* probabilities ($p = 0.0425$, $p = 0.1024$, etc.) for the outcome of significance tests. If the package doesn't tell you whether the exact probability it quotes is significant, simply apply the threshold value rule as before. Thus, on the 5 per cent criterion, $p = 0.0425$ is significant, because it is less than 0.05, but $p = 0.1024$ is not, because it is greater than 0.05.

An important point must be made here regarding the inference to be drawn from achieving different levels (10, 5, 1 per cent, etc.) of significance. A high level of significance is not the same as a large effect in the sense of a large difference or a steep trend. The magnitude of an effect – difference or trend – is usually known as an *effect size*. This is quite different from the level of significance. The distinction is made clear in Fig. 3.2. The figure shows two trends. In Fig. 3.2a,

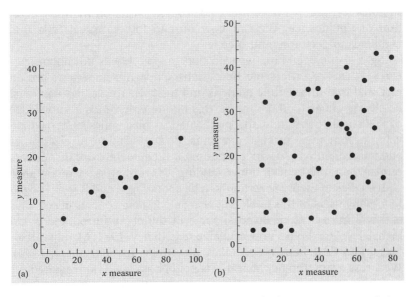

Figure 3.2   Scattergrams showing the effect of sample size on apparent trends (*see text*).

the *y* measure increases in close relationship with the *x* measure, yielding what looks like a clear positive trend. Figure 3.2b, on the other hand, shows a scatter of points in which it is more difficult to discern a trend. However, if we perform a suitable significance test for a trend on the two sets of data, the trend in Fig. 3.2b turns out to be significant at the 1 per cent level while that in Fig. 3.2a isn't even significant at the 10 per cent level. The crucial difference between the two trends, of course, is the sample size. The number of data points in Fig. 3.2a is low, so a few inconsistencies in the trend are enough to push it below significance. Figure 3.2b, however, has a large number of points; so even though there is a wide scatter, the trend is still significant. Exactly the same sample size effect would operate in the case of difference analyses.

Because the level of significance by itself gives little indication of the magnitude of a difference or trend, it is always helpful to provide such an indication, usually in the form of summary statistics and their associated sample sizes. We shall return to this point later.

3.3   Significance tests

So far, we have seen how to get round the problem of subjective impression in interpreting data by using the criterion of statistical significance, and we have looked at some caveats on the interpretation of significance. We can now turn to the statistical *tests* that enable us to decide significance.

Right at the beginning we said that this book was not about statistics. It isn't. At the same time, statistical significance tests are an essential tool in scientific analysis and the rules for using them must be clearly understood. However, this no more demands a knowledge of statistical theory and the mathematical mechanics of tests than using a computer package demands an appreciation of electronics and microcircuitry. As with most tools, it is competent *use* that counts rather than theoretical understanding. Nevertheless, acquaintance with statistical theory is certainly to be encouraged and it is envisaged that many users of this book will also be pursuing courses in statistics and will have at their disposal some of the many introductory and higher-level textbooks now available (e.g. Bailey, 1981; Meddis, 1984; Siegel and Castellan, 1988; Sokal and Rohlf, 1995; Dytham, 1999). We stress again, though, that this is neither assumed nor required for our purposes. Our aim here is simply to introduce the use of significance testing as a basic tool of enquiry and some simple, but broadly applicable, tests for differences and trends.

3.3.1   Types of measurement and types of test

The test we choose in a particular case may depend on a number of things. The following are three important ones.

1   Types of measurement

The first is the kind of measurement we employ. Without getting too bogged down in jargon, we can recognize three kinds.

Nominal or classificatory measurement.   Here, observations or recordings are allocated to one of a number of discrete, mutually exclusive categories such as male/female, mature/immature, red/yellow/green/ blue, etc. Thus if we were to watch chicks pecking at red (R), green (G) and orange (O) grains of rice and recorded the sequence of pecks with respect to the colour targeted, we might end up with a string of data as follows:

R   O   R   R   G   G   O   R   G   G   G   G   O   O   G   R   O

Such data are measured purely at the level of the category to which they belong and measurement is thus *nominal* or *classificatory*.

Ordinal or ranking measurement.   In some cases, it may be desirable (or necessary) to make measurements that can be *ranked* along some kind of scale. For instance, the intensity of the colour of a turkey's wattles might be used as a guide to its state of health. The degree of redness of the wattles of different birds could be scored on a scale of 1 (pale pink) to 10 (deep red). The allocation of scores to wattles is arbitrary and there is no reason to suppose that the degree of redness increases by the same amount with each increase in score. Thus the difference in redness between scores of 8 and 9 might be greater than the difference between scores of 2 and 3. All that matters is that 9 is redder than 8 and 3 is redder than 2; the absolute difference between them cannot be quantified meaningfully.

Constant interval measurements.   In other scale measurements, the difference between scores can be quantified so that the difference between scores of 2 and 3 is the same as that between scores of 8 and 9. Such measurements may have arbitrarily set (e.g. scales of temperature) or true (e.g. scales of time, weight, length) zero points. Such *constant interval measurements* can in fact be split into two categories on the basis of arbitrary versus true zero points and their scaling properties (e.g. Martin and Bateson, 1993), but this is not important here.

While defining measurements seems rather dry and theoretical, we need to be aware of the kind of measurement we use because, as we shall see, some significance tests are very restrictive about the form of data they can accept. Another reason for highlighting it, is that we should always seek the measures that give us the maximum amount of information. Usually this means constant interval measurements, because they are on a continuous, non-arbitrary scale. However, nominal or ordinal data are sometimes more appropriate and not infrequently the best that can be achieved.

2 Parametric and nonparametric significance tests

The second thing we must keep an eye on is the nature of the data set itself, in particular the sample size and the distribution of values within the sample. Again, detailed consideration of this is unnecessary but it is a factor that determines the range of tests we shall be introducing so a brief discussion is warranted.

Parametric tests. These make a number of important assumptions that are frequently violated by the kinds of data sets collected during practical exercises. The most critical concerns the distribution of values within samples. Parametric tests generally assume that the data conform (reasonably closely at least) to what is known as a *normal* distribution.

As Dytham (1999) puts it, the normal distribution is the most important distribution in statistics (*but see* Box 3.1) and it is often assumed (all too frequently without checking) that data are distributed in this way. We've already encountered it in our discussion of frequency distributions and it is illustrated again in Fig. 3.3. Essentially a normal distribution demands that most of the data values fall in the middle of the range (cluster about the mean) with the number tapering off symmetrically either side of the mean to a few extreme values in each of the two tails. The height of the adult male or female population in a city would look something like this: most people would be around the average height for their sex, some would be quite tall or quite short and a few would be extremely tall or extremely short. While normality is not the only assumption underlying parametric tests, the arithmetic of such tests is based on the parameters describing this symmetrical, bell-shaped curve (hence the term *parametric*). The more distorted (less normal) the distribution becomes, therefore, the less meaning the calculations of parametric tests have. We can convince ourselves of this by briefly considering some basic features of the distribution. We shall then consider what to do if the data are not normal.

The standard deviation and probability. We've talked about the arithmetic mean as a measure of central tendency. This is a useful

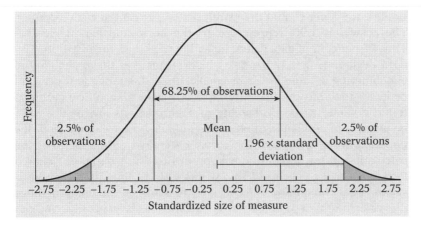

Figure 3.3 A standardized normal distribution (with a mean of 0) and showing one standard deviation (vertical lines) either side of the mean (modified from Dytham, 1999).

parameter that tells us one thing about the nature of the data set. What it doesn't tell us, of course, is anything about the *variation* in the data. This is where we come back to the *standard deviation*, first encountered in calculating the standard error to the mean in Box 2.1. The standard deviation (*see* calculation in Box 2.1) is a measure of the spread of data values about the mean, but instead of simply being the full range of *actual* values from smallest to largest (which will vary with sample size), it reflects the theoretical spread of the majority of values (68.25 per cent of them in a perfectly normal distribution) in the true population. We can represent this as two vertical lines (one either side of the mean) in Fig. 3.3, which is actually a standardized normal distribution (where the mean is subtracted from each data value and the result divided by the standard deviation, thus giving a mean of zero and a variance of 1). An easy way to visualize the standard deviation is as the point of inflection either side of the mean (where the curve of the normal distribution changes from convex to concave). Taking the majority spread like this avoids the distortion that might be inflicted by odd outlier values at the extremes and provides a convenient standard yardstick of variability within data sets. Moving two or three standard deviations away from the mean includes predictably greater percentages of the data set: 94.45 per cent in the case of two standard deviations and 99.97 per cent in the case of three. The important point here is that we can use this property when it comes to significance testing. Since we know what percentage of the data points is included within different multiples of the standard deviation, we can easily work out how many standard deviations would include 95 per cent of the data or, more to the point, exclude the 5 per cent at the extremes (2.5 per cent at each extreme). The answer is 1.96, represented by the horizontal bar in Fig. 3.3. Thus 95 per cent of data values lie within the range 'mean ± 1.96 standard deviations ($\bar{x} \pm 1.96$ s.d.)', which means that the probability of encountering a value larger or smaller than this range is 5 per cent or less – the conventional threshold of statistical significance! If we found a value with a probability of occurrence as low as this we'd be justified in concluding that it was unlikely to have come from the population that generated the curve. Of course, as the fact that the calculation of the standard deviation in Box 2.1 is based on squared deviations makes clear, all this works *only* as long as the curve is symmetrical and conforms respectably to a normal distribution. Since the calculation of parametric significance tests generally employs the same squared deviations procedure, as we shall see shortly, the restrictions apply to all these tests.

One additional point. Quite a number of significance tests are based on calculating the deviation of an observed mean from the null expectation of a standardized normal distribution (Fig. 3.3). Conventionally this is called a z-value and, from the above, we can see that if z exceeds 1.96 the result is significant at the 5 per cent level. (The value 1.96 is actually the threshold value when both sides – 'tails' in the jargon – of the distribution are taken into account, a so-called two-tailed test; when only one tail is considered (a one-tailed test), the threshold value is 1.64. We shall discuss one- and two-tailed tests in detail later.)

Departures from normality. Frequency distributions can depart from normality in a number of ways. Two broad kinds of departure, however, are *skewness* and *kurtosis*. Skewness is a synonym for asymmetry, i.e. one or other tail of the distribution is more drawn out than the other. The distribution in Fig. 2.7b is said to be skewed to the right (towards the y-axis), while the opposite bias would be skewed to the left. Kurtosis refers to the flatness of the distribution, which can be *leptokurtic* (more values are concentrated around the mean and in the tails and fewer in the 'shoulders' of the distribution), or *platykurtic* (where the reverse is true). Bimodal distributions, like that in Fig. 2.7c, are thus extremely platykurtic.

Percentages and proportions present their own problems for normality. Because they range between 0 and 100, or 0 and 1, the distribution of values is artificially truncated at either end. This may not present a serious problem if most of the values in the data set occur in the middle two-thirds or so of the distribution, but if they approach 0 or 100/1 there is cause for concern.

Transformations. So what do we do if we suspect our data may not be normally distributed? Happily, and as long as our sample size is big enough (> 50 as a rough guide) to make a comparison meaningful, the wide range of statistical packages now available for personal computers makes the answer simple. Test it! There are some well-established significance tests (such as the chi-squared and Kolmogorov–Smirnov tests) that allow comparisons between frequency distributions of data and various theoretical distributions of which the normal is the commonest. We shall come to these when we have said a little more about significance tests generally. If we test our distribution and find it does not differ significantly from normal, then we're at liberty to use any appropriate parametric test at our disposal. If it does differ, we can do one of two things. We can abandon the idea of using parametric statistics and choose an appropriate *nonparametric* test instead (*see below*), or we can *transform* the data to see whether it can be normalized. Several transformations are available but the most widely used are probably *logarithmic* (log x or, where there are zeros in the untransformed data, $\log(x + 1)$) or *square root* transformations. Simply log or take the square root of each data value, recast the distribution and test it for normality again. Where percentages or proportions stray below about 30 per cent (or 0.3) or above 70 per cent (or 0.7), an *arcsine square root* transformation (calculated by taking the square root of the proportion – so divide percentages by 100 first – then the inverse sine (sin^{-1} on many calculators) of the result) will stretch out the truncated tails and prevent undue violation of the normality assumption of parametric tests.

Of course, even transformation may not succeed in normalizing our data, in which case we must seriously consider using nonparametric statistics. Indeed, we may not even get as far as worrying about the normality of our data before opting for a nonparametric approach. Among their various other requirements, parametric tests demand that measurements are of the constant interval kind, so cannot usually deal with the other types of measurement we might be forced to use. Nonparametric tests are much less restrictive here.

Nonparametric tests.   Nonparametric tests are sometimes referred to as distribution-free, *ranked* or *ranking* tests because they do not rely on data's being distributed normally and generally work on the ranks of the data values rather than the data values themselves. While they may be distribution-free, however, they are not entirely assumption-free. They assume the data have *some* basic properties, such as independence of measurement (*see later*) and a degree of underlying continuity (*see* Martin and Bateson, 1993), but in most cases these assumptions are easily met. In the jargon of statisticians, nonparametric tests are thus more *robust* because they are capable of dealing with a much wider range of data sets than their parametric equivalents. While they can deal with the same constant interval measurements as parametric tests, they can also cope with ordinal (ranking) and classificatory measurements. Nonparametric tests are especially useful when sample sizes are small and assumptions about underlying normality particularly troublesome. There are a couple of drawbacks, however. The first, arguably overstated (*see* Martin and Bateson, 1993), weakness is that nonparametric tests are generally slightly less powerful (*power* here meaning the probability of properly rejecting the null hypothesis – we shall return to this shortly) than their parametric equivalents. The second, which is slowly being addressed (*see for example* Meddis, 1984), is that the range of tests for more complex analyses involving several variables at the same time is very limited. Sophisticated multivariate analysis is still the undisputed province of parametric statistics. Nevertheless, for our purposes, and with one exception, there are perfectly good parametric and nonparametric equivalents and we shall introduce them side by side in our discussion of significance testing.

Box 3.1	Types of distribution

The normal (or Gaussian) distribution is a form of *continuous* distribution, where, in principle, data can take a continuum of values. There are other forms of continuous distribution, but the normal is by far the most relevant to biological situations. However, there is another family of distributions which is important in biology. These are *discontinuous* or *discrete* distributions: two are particularly relevant.

The Poisson distribution

The Poisson distribution describes the occurrence in units of time or space, for instance the number of solitary bee burrows scored in a quadrat or the number of hedgehogs dropping into a cattlegrid overnight. The key assumptions are: (a) the mean number of occurrences per unit is small in relation to the maximum number possible – i.e. occurrences are rare, (b) occurrences are independent of each other – i.e. there is no influence of one on the likelihood of another, (c) occurrences are random. Indeed, the reason a Poisson distribution is fitted to data is to test for independence or randomness in time or space. Poisson data are characterized by the mean's being equal to the square of the standard deviation (known as the *variance*). If the variance is bigger than the mean, the data are more clustered than random; if it is smaller, they tend towards uniformity. A glance at these two summary statistics thus gives a good indication of the kind of distribution we're dealing with.

The binomial distribution

The binomial distribution is a discrete distribution of occurrences where there are two possible outcomes for each so that the probability of one outcome determines the probability of the other. Thus if the probability of an egg's hatching is 0.75, the probability that it won't hatch is 1 − 0.75, i.e. 0.25. This logic can be extended to calculate the probability of various combinations of hatching (H) and failed (F) eggs in clutches of different size. Thus if two eggs are laid, there are four possible outcomes: HH, HF, FH, FF. Applying the values above gives a probability of 0.56 (0.75 × 0.75) for both eggs, hatching successfully, 0.19 (0.75 × 0.25) for each of the two mixed outcomes (HF and FH) and 0.06 for two failures. The same could be done for clutches of three, four or however many eggs. The binomial distribution thus gives a baseline chance probability against which outcomes in various situations can be judged. So, if we found a population in which the proportion of two-egg clutches failing completely was 30 per cent instead of 6 per cent, we might become suspicious about the health of the birds or their environment.

3   One-tailed versus two-tailed, and general versus specific tests

The third important factor we must consider relates to the prediction we are trying to test. Suppose we are predicting a difference between two sets of data, say a difference in the rate of growth of a bacterial culture on agar medium containing two different nutrients. We could

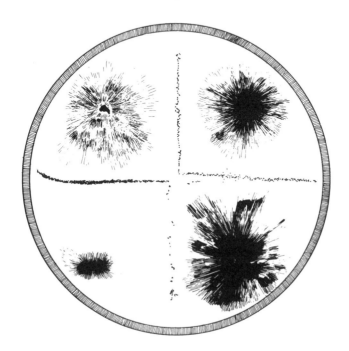

make two kinds of prediction. On the one hand, we could predict a difference without implying anything about which culture should grow faster. In this case, we wouldn't care whether culture A grew faster than culture B or vice versa. This is a *general* prediction. On the other hand, we might predict that one particular culture would grow faster than the other, e.g. A would grow faster than B; this is a *specific* prediction. Which of these kinds of prediction we make affects the way we test the predictions.

The same distinction arises with trend predictions. Imagine we want to know whether there is a trend between the size of a male cricket and the number of fights he wins over the course of a day. We can make a general prediction (there will be a trend, positive or negative), or we can make a specific prediction (larger males will win more fights; i.e. the trend will be positive). We can think of the general prediction as incorporating both positive and negative trends; either would be interesting. The specific prediction is concerned with only one of these.

In cases like those above, where there are only two possible specific predictions within the general one, we can use the same significance test for either general or specific predictions but with different threshold probability levels for the test statistic (*see below*) to be significant. Because here the specific and general predictions are concerned with one and two directions of effect respectively, the threshold value of the test statistic at the 5 per cent level in the general test becomes the threshold value at the 10 per cent level in the specific test. In statisticians' jargon, we thus do either a one-tailed (specific) or a two-tailed (general) version of the same test. There is, of course, an obvious, and dangerous, trap for the unwary here. The trap is this: if the value of a test statistic just fails to meet the 5 per cent threshold in a two-tailed test, there is a sore temptation to pretend the test is really one-tailed so that the test statistic value appropriate to the 10 per cent probability level in a two-tailed test can be used, thus increasing the likelihood of achieving a significant result. *It must be stressed that this is tantamount to cheating. A one-tailed test is legitimate only when the prediction is made in advance of obtaining the result **and** when results in the opposite direction can reasonably be regarded as equivalent to no difference or trend at all. It is completely inadmissible as a fallback when a two-tailed test fails to yield a significant outcome.* A one-tailed test should thus be used only when there are genuine reasons for predicting the direction of a difference or trend *in advance*.

However, the distinction between general and specific tests is not simply that between two-tailed and one-tailed tests. Instead of the above, for instance, imagine that we are predicting a difference between three (or more) groups, say weight of fruit produced in a season from trees treated in three different ways: A, B and C. As before, we can make the general prediction that there will be a difference between the three treatments but, unlike before, this general prediction incorporates *six* potential specific predictions about how the treatments will differ:

$$A > B > C, \quad B > C > A, \quad C > A > B, \quad A > C > B, \quad B > A > C, \quad C > B > A$$

This extra complexity means that we have to use different tests for a general as opposed to a specific prediction – we shall see how later – and cannot simply use one-tailed and two-tailed versions of the same test. The one-tailed/two-tailed distinction is thus a special case of the difference between general and specific tests.

3.3.2   Simple significance tests for differences and trends

Having discussed the general principle of statistical significance, we come now to some actual tests which allow us to see whether differences or trends are significant at an appropriate level of probability. A glance at any comprehensive statistics textbook will reveal a plethora of significance tests for both kinds of analysis. These cater for the various subtleties of assumption and requirement for statistical power under different circumstances. Many of these tests, however, are sophistications of more basic tests which are suitable for a wide range of analyses. Here, we introduce a selection of such basic tests which can be used with most kinds of data that are likely to be collected in practical exercises. Where appropriate, parametric and nonparametric equivalents are presented side by side. Tests are presented in outline in the text but detailed examples of calculations are given in Appendix II. The tests can thus be calculated by hand if necessary.

Test statistic

Significance tests calculate a test statistic that is usually denoted by a letter or symbol: t, H, F, χ^2, r, r_s and U are a few familiar examples from various parametric and nonparametric tests. The value of a test statistic has a known probability of occurring by chance for any given sample size or what are known as *degrees of freedom* (*see later*). A calculated value can thus be checked to see whether it concurs with or passes (positively or negatively, depending on the test) the threshold value appropriate to the level of probability chosen for significance. This usually means comparing the value with a table of threshold values. Such comparisons are made automatically for the tests in many statistical computer packages.

3.3.3   Tests for a difference

We shall introduce three types of difference test, with parametric and nonparametric equivalents presented together: first χ^2 (chi-squared, pronounced 'ky-squared'), then the t-test and Mann–Whitney U-test as parametric and nonparametric (respectively) tests for a difference between two groups, and finally analysis of variance (parametric and nonparametric versions). All test predictions of difference, but each deals with a different kind of measurement or number of groups for comparison. In fact t- and U-tests are really analyses of variance in themselves, but they are specifically designed to compare two groups of data and, as we shall see, cannot be used when more than two groups are being compared.

Because parametric tests of significance are all based on the normal distribution, they share a set of core calculations, known as the *sum of squares*. We have encountered these in part already in calculating the standard deviation and standard error (Box 2.1). To avoid unnecessary repetition, we present these common core calculations separately in Box 3.2 and refer back to them in subsequent calculations of significance tests. It is also important to point out that many parametric, and nonparametric, tests of difference assume the variances (scatter in the data, *see* Box 2.1) of the groups being compared are the same. This requirement should be checked and the data, if need be, tested appropriately for *homogeneity of variances*. Bailey (1981) provides a clear guide.

Box 3.2	**The basic calculations for parametric test statistics**

If you calculate the test statistic of many parametric statistical tests by hand, the calculations often involve a standard set of core operations on the data. Rather than repeating them in all the relevant boxes, we have placed them here for reference. They all involve calculating a quantity known as the *sum of squares*, which is actually the same operation performed when finding the standard deviation or standard error (*see* Box 2.1).

1. Follow steps 1–3 of Box 2.1. The resulting quantity is the *sum of squares of x*, conventionally denoted by S_{xx}.

Sometimes you will have a y-variable as well as an x-variable (e.g. in regression, *see* p. 82). If so:

2. Calculate the sum of all the y-values in the data set (Σy).

3. Square all the y-values and sum them, giving (Σy^2).

4. Multiply the x- and y-values of each x–y pair together, and add them together, giving (Σxy).

5. Calculate $\Sigma y^2 - [(\Sigma y)^2/n]$, giving the *sum of squares of* y, S_{yy}.

6. Calculate $\Sigma xy - [(\Sigma x)(\Sigma y)/n]$ giving the *sum of the cross-products*, S_{xy}.

Tests for a difference between two groups

We shall start with the relatively simple situation of comparing two groups. Here, two mutually exclusive groups (e.g. male/female, small/large, with property a/without property a, etc.) have been identified and measurements made with respect to each (e.g. the body length of males versus the body length of females, the number of seeds set by small plants versus the number set by large plants, the survival rate of mice on drug A versus the survival rate on drug B). Depending on the kind of measurement made, we can use one of a number of tests to see whether any differences are significant.

Chi-squared (χ^2). A chi-squared test can be used if data are in the form of counts, i.e. if two groups have been identified and observations classified in terms of the number belonging to each. Chi-squared can be used *only* on raw counts; it cannot be used on measurements (e.g. length, time, weight, volume) or proportions, percentages, or any other derived values. The test works by comparing observed counts with those expected by chance or on some prior basis. As an example, we can consider a simple experiment in Mendelian inheritance. Suppose we crossed two pea plants that are heterozygous for yellow and green seed colour, with yellow being dominant. Our expectation from the principles of simple Mendelian inheritance, of course, is that the progeny will exhibit a seed colour ratio of 3 yellow : 1 green. We can use the chi-squared test to see whether our observed numbers of yellow and green seeds differ from those expected on a 3 : 1 ratio. The expected numbers are simply the total observed number of progeny divided into a 3 : 1 ratio. Thus:

	Seed colour		
	Yellow	Green	Total
Number observed	130	46	176
Number expected	132	44	176

To find our χ^2 test statistic we calculate:

$$\chi^2 = \Sigma(O - E)^2/E$$

where O is the observed number and E the expected number in each group. In our example, therefore:

$$\chi^2 = (130 - 132)^2/132 + (46 - 44)^2/44$$

$$= 0.0303 + 0.0909$$

$$= 0.1212$$

We can now check our calculated value of χ^2 against the threshold values in Appendix III, Table A. To do this, however, we must first decide on the appropriate number of *degrees of freedom* to use. Degrees of freedom (often referred to simply as d.f.) are related to sample size, but taking into account the number of groups or classes into which the data are cast for analysis and the consequent reduction in independence between data values. Essentially, the more *operations* performed within an analysis, the fewer the available degrees of freedom for interpreting the eventual outcome. Exactly how the appropriate degrees of freedom are calculated in any particular test will therefore depend on how the data are handled. In the two-group chi-squared test above, the number of degrees of freedom is equal to the number of groups minus one (i.e. $2 - 1 = 1$). The threshold χ^2 value in which we are interested is therefore the value appropriate to a 5 per cent (0.05) probability for one degree of freedom. Reference to Appendix III, Table A, shows this value to be 3.84. To be significant at the 5 per cent level, our calculated χ^2 value must *equal or exceed* 3.84. Since 0.1212 is less than 3.84, the null hypothesis that there is no departure from a 3 : 1 ratio cannot be rejected and there is no reason to suppose that seed colour is segregating in anything other than a simple Mendelian fashion. If we had set our threshold of significance at 1 per cent (0.01) instead of 5 per cent, the value that χ^2 would have had to equal or exceed is 6.63 (Table A). The criterion thus becomes tougher the smaller the margin of error we impose.

In the pea example, the expected numbers were dictated by the Mendelian theory of inheritance; we have good reason to expect a 3 : 1 ratio of yellow : green seeds and thus a difference between groups. In many cases, of course, we should have no particular reason for expecting a difference and our expected numbers for the two groups would be the same (half the total number of observations each). Thus if our yellow and green groups had referred to pecks by chicks at one of two different coloured grains of rice on a standard

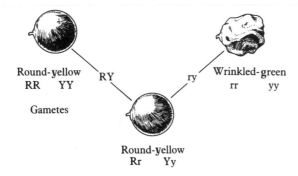

Round-yellow
RR   YY

RY

ry

Wrinkled-green
rr   yy

Gametes

Round-yellow
Rr   Yy

background instead of the inheritance of seed colour, our chi-squared table would have looked very different:

	Grain colour	
	Yellow	Green
Number of pecks observed	130	46
Number of pecks expected	88	88

and the result would have been a χ^2 value of 40.09, which exceeds even the threshold value (10.83) for a significance level of 0.1 per cent (0.001) (*see* Appendix III, Table A). In this case we could safely reject the null hypothesis of no difference in the number of pecks to different coloured grains and infer a bias towards yellow on the part of the chicks.

An important point to bear in mind with χ^2 is that it is not very reliable when small samples are tested. It is a good idea not to use it when the sample size is smaller than 20 data values or when any expected value is less than 5.

t-*tests* and *Mann–Whitney* U-*tests*. *t*-tests and Mann–Whitney *U*-tests can both be performed on raw counts like chi-squared, except that they deal with each contributing data value in the two groups separately instead of as a single total. Thus if the pecking data in our last example of chi-squared were derived from ten chicks given the opportunity to peck at yellow grains and ten given the opportunity to peck at green grains on a standard background, the values that would be used in the chi-squared and the *t*- or *U*-tests can be indicated as follows:

Chick	Pecks to yellow	Chick	Pecks to green	
a	12	k	2	
b	14	l	3	
c	13	m	10	
d	3	n	6	
e	23	o	4	A t- or U-test uses these values
f	13	p	5	
g	11	q	3	
h	15	r	1	
i	9	s	7	
j	17	t	5	
Total	130		46	A chi-squared test uses these values (or, in principle, any subtotal of data values)

In addition, however, t- and U-tests can deal with data other than counts. We can use them to compare two groups for constant interval data such as body size, time spent performing a particular behaviour, percentage of patients responding to different drug treatments or a host of other kinds of data. The nonparametric U-test can also cope with ordinal (rank) data. In addition, unlike some kinds of two-group test working on individual data values, the U-test and most forms of the t-test do not require equal sample sizes in the two groups. However, and we stress this again, these tests can be used *only* for comparing *two* groups. Furthermore, the U-test can only test a general (or two-tailed) prediction (i.e. there is a difference, but not in any one predicted direction). Calculation of the test statistics t and U is straightforward and is shown in Box 3.3a, b.

Box 3.3a

(i) Mean values: how to do a general parametric test for two groups (two-tailed t-test)

1. Frame the prediction. In this case it is the general prediction that the two groups (A and B) will have different mean values (A ≠ B). Thus the null hypothesis is that the two group means will not differ.

2. Count the number of data values in the first group; this number is referred to as n_1. If $n_1 = 1$, then there is definitely something wrong! (If this number represents a count of the number of items in one of two categories – that form the two groups – then you should be doing a χ^2 test. If it is a rank or a constant-interval measurement, then you need to collect some more data!) Count the number of data values in

the second group; this number is referred to as n_2 (it should be greater than 1, as before).

3. Calculate the variances of each group separately (*see* Box 2.1 for how to do this), producing s_1^2 and s_2^2.

4. Calculate the value of $P = (n_1 + n_2)/(n_1 n_2)$.

5. Calculate the value of $Q = (s_1^2(n_1 - 1) + s_2^2(n_2 - 1))/(n_1 + n_2 - 2)$.

6. Calculate the value of $R = \sqrt{(PQ)}$. This is the standard error of the difference between the two groups.

7. Calculate the mean values of each group separately, and take the difference $S = (\mu_1 - \mu_2)$. Switch the mean values round so as to make the difference positive, since in a general test we are not interested in the direction of the difference, but merely in whether it differs from zero.

8. Calculate the value of the test statistic, $t = S/R$.

9. In order to calculate t, we needed to know the difference in mean values, and the s.e. of this difference, i.e. two prior parameters were required. The degrees of freedom of t are therefore $n_1 + n_2 - 2$.

10. Look up the two-tailed value of t in Table D of Appendix III for the critical value for your degrees of freedom. If your value is greater than the relevant value in the table, then the difference you found is significant.

A worked example of a general t-test can be found in Appendix II.

(ii) Mean values: how to do a specific parametric test for two groups (one-tailed t-test)

1. Frame the prediction, i.e. decide which of the two groups (A and B) is predicted to have the greater mean value. There needs to be some a priori reason (theory, or previous published or gathered data) for this prediction. Suppose that on the basis of your knowledge, you predict that $A > B$. The null hypothesis is that the mean values for the two groups do not differ in the predicted direction.

2. Calculate the value of t as steps 2–9 above, but make sure that the difference in mean values is done in the way that is *predicted* to generate a positive difference. Here you are predicting that $A > B$, and hence $(A - B)$ should generate a positive value of t. If it does not actually generate a positive difference, then you know automatically that the result is not significant. Note that if the result is an unusually large but negative value of t (that would have been significant, had you predicted the opposite pattern of mean values), you are not allowed to conclude *anything* other than that your prediction is not supported by the data. This

is the cost of a specific (one-tailed) prediction paid in exchange for the benefit of the more powerful test.

3. Look up the critical value of a one-tailed t-test in Table D of Appendix III using the degrees of freedom you have. If your t-value is greater than the critical one, then you conclude that the result is significant: the evidence suggests that the group mean predicted to be greater really is so, and you reject the null hypothesis.

A worked example of a specific t-test can be found in Appendix II.

Box 3.3b

(i) Mean ranks: how to do a general nonparametric test for two groups (Mann–Whitney U-test)

1. Count the number of data values in the group with the fewer values (if there is one); this number is referred to as n_1.

2. Count the number of data values in the other group; this is n_2.

3. Rank all the values *in both groups combined*. The smallest value takes the lowest rank of 1, the next smallest value a rank of 2 and so on. If two or more values are the same they are called *tied values* and each takes the average of the ranks they would otherwise have occupied. Thus, suppose we have allocated ranks 1, 2 and 3 and then come to three identical data values. If these had all been different they would have become ranks 4, 5 and 6. Because they are tied, however, they each take the same average rank of 5 ($(4 + 5 + 6)/3$), though – and this is important – the next highest value still becomes rank 7 just as if the three tied values had been ranked separately. If there had been only two tied values they would each have taken the rank 4.5 ($(4 + 5)/2$) and the next rank up would have been 6. We should thus end up with rank values ranging from 1 to N, where $N = n_1 + n_2$.

4. Add up the rank values within each group giving the total R_1 and R_2 respectively.

5. Calculate $U_1 = n_1 \times n_2 + ((n_1(n_1 + 1))/2) - R_1$.

6. Calculate $U_2 = n_1 \times n_2 - U_1$.
   If U_2 is smaller than U_1 then it is taken as the test statistic U. If not, then U_1 is taken as U.

7. We can now check our value of U against the threshold values in U tables (a sample is given in Appendix III, Table B). If our value is *less* than the threshold value for a probability of 0.05, we can reject the null hypothesis that there is no difference between the groups. Note that, in this test, we use the two sample sizes, n_1 and n_2, rather than degrees of freedom to determine our threshold value.

If one of the groups has more than 20 data values in it, U cannot be checked against the tables directly. Instead, we must use it to calculate another test statistic, z, and then look this up (some sample threshold values are given in Appendix III, Table C). The calculation is simple:

$$z = \frac{U - (n_1 \times n_2)/2}{[(n_1)(n_2)(n_1 + n_2 + 1)]/12}$$

A worked example of a U-test can be found in Appendix II.

(ii) Mean ranks: how to do a specific nonparametric test for two groups

There is no nonparametric test specially designed to test for a specific difference between two groups. Use the specific nonparametric analysis of variance (*see* Box 3.4a), because this can cope with any number of groups.

Tests for a difference between two or more groups

So far, we have introduced significance tests that can test for a difference between two groups. In many cases, of course, we shall be faced with more than two groups. For instance, Fig. 2.4 suggests that the amount of damage around the edges of leaves increases with leaf size class, perhaps indicating a predilection for big leaves by particular kinds of pest. If we wanted to know whether the difference between the size classes was significant, we should have to deal with three groups of data. How do we do it? The temptation to which many succumb is to do a *round robin* comparison of pairs of groups using a *t*- or *U*-test. In our Fig. 2.4 example, this would mean testing for a difference between small and medium leaves, then for a difference between medium and large leaves, and finally for a difference between small and large leaves. *The error of this cannot be emphasized too strongly.* The most serious problem arising from such a practice is that it increases the likelihood of obtaining a significant difference by chance when really none exists. To take an extreme example: if we carried out 100 two-group comparisons, then, just by chance, five of them stand to be significant at the 5 per cent level. Even if we made only 20 comparisons, one is likely to be significant by chance. While this may not seem a serious difficulty when we are dealing with only three or four groups, these examples illustrate the error in principle. To get round the problem, we need tests that can cope with comparisons between several groups at the same time.

One-way analysis of variance.   Where we have series of data values falling into several groups (in a similar fashion to data values in the two groups of a *t*- or *U*-test), a one-way analysis of variance (often expressed as the acronym *one-way ANOVA*) is a suitable significance test. There are both parametric analyses of variance and nonparametric tests that have an equivalent function. We shall introduce both kinds here. While parametric analyses of variance are by far the more widely used, the nonparametric version we shall describe has the advantage not only of being robust to a wider range of data but also of allowing specific predictions about the *direction* of differences between groups to be tested. This test can thus be used to test general or specific predictions for two or more groups. For two groups it is therefore more flexible than the *U*-test and, as a result, we recommend it even in these cases. The two kinds of analysis of variance are outlined in Box 3.4a, b.

Box 3.4a	**(i) Mean values: how to do a general parametric one-way analysis of variance (ANOVA)**

1. Frame the prediction. In this case, the general prediction being made is to ask whether there are any differences among the mean values of the groups. Therefore the null hypothesis actually tested is that there are no differences among the mean values of the groups.

2. The test considers i groups of data, each of which contains n_i data values. The total number of data values in all groups together $= N = \Sigma n_i$.

3. Calculate the mean values of each of the groups, μ_i.
   Work out T, the total sum of squares of all the data (follow Box 3.2, items 1–3).

4. Work out S_i, the sum of squares for the data of each group separately (again, follow Box 3.3, 1–3).

5. Calculate the error sum of squares, $SS_{error} = \Sigma S_i$.

6. Calculate the among-groups sum of squares, $SS_{among} = T - SS_{error}$. This should give the same result as calculating the sum of squares using the mean values μ_i rather than the raw data.

7. The d.f.$_{total}$, the total degrees of freedom, is $N - 1$.

8. The among-groups degrees of freedom, d.f.$_{among}$, is $(i - 1)$.

9. The error degrees of freedom, d.f.$_{error}$, is $(N - i)$.

10. The mean square among groups, $MS_{among} = SS_{among}/\text{d.f.}_{among}$.

11. The error mean square, $MS_{error} = SS_{error}/\text{d.f.}_{error}$.

12. The test statistic, $F = MS_{among}/MS_{error}$.

13. The degrees of freedom for F are $\text{d.f.}_{\text{among}}$, $\text{d.f.}_{\text{error}}$ (note that F has *two* values for degrees of freedom, unlike other test statistics, which have only one).

14. Now look up the critical value of F in Table G of Appendix III for $f_1 = \text{d.f.}_{\text{among}}$, and $f_2 = \text{d.f.}_{\text{error}}$.

15. If your value for F is greater than the critical value, then the result is taken to be significant, and we reject the null hypothesis of equal mean values for all groups. Present the one-way ANOVA laid out in the standard manner (*see* worked example in Appendix II).

16. Note that this is a *general* prediction and, therefore, we can only conclude that there are *some* differences among the means: we are *not* allowed to say which particular pair of mean values are significantly different because we did not make any specific prediction beforehand about the ordering of the mean values, nor did we set up planned specific contrasts between particular sets of groups. Many researchers would like to be able to make such *post hoc* tests, however, and a plethora of different methods are available and are often used. However, they should only be used as a rough guide to what the differences might be, preferably to set up a priori hypotheses for a new data set to test. Many people use the least significant difference, or LSD test. All this does is to take the MS_{error} from the ANOVA (which is an estimate of the variation within groups) and construct a 95 per cent confidence limit with it using:

$$LSD = t_{N-i} \times \sqrt{[(2/n) \times MS_{\text{error}}]}$$

where t_{N-i} is the 5 per cent threshold value of t for $N - i$ degrees of freedom. This looks a simple enough procedure, but in fact the LSD test is only accurate for a priori contrasts. It should, therefore, only be used as a rough guide in *post hoc* testing: *see* p. 70.

(ii) Mean values: how to do a specific parametric one-way ANOVA

Unfortunately there is effectively no parametric equivalent of the test for a particular rank order of mean ranks, as in Box 3.4b(ii). Although one does exist, called isotonic regression (*see* Gaines and Rice, 1990), it is very obscure and hardly ever used; it also involves computer-intensive randomizations of the data, and is well beyond the scope of this book. More common (but still unusual) is the use of a priori contrasts. If done in advance of obtaining the data, these allow $(i - 1)$ contrasts to be made. We explain this technique here.

1. Each contrast consists of one subset of the i groups (A) contrasted against another subset (B) (for example, a control (A) versus all treated groups (B)): each therefore involves effectively creating *two* groups out

of the data, and testing them using a t-test. Such contrasts can them-selves be either general (A ≠ B) or specific (e.g. A > B).

2. Frame the predictions. Here you do this by formulating the contrasts that you want to make *in advance of collecting the data*. You are allowed $(i-1)$ contrasts. Express each contrast as an inequality, and make the left-hand side greater or equal to zero. For example, if you have three groups, (A, B and C) and want to test whether group A is different from the other two groups (i.e. from the average of B + C), then this general hypothesis is that:

$$A \neq (B + C)/2$$

and hence

$$2A - B - C \neq 0, \text{ i.e. } +2(A) - 1(B) - 1(C) \neq 0$$

This gives us a set of coefficients to apply to the mean values of each group; in this case they are +2, –1 and –1 for the mean values of groups A, B and C, respectively.

Groups that you want to leave out of the contrast have a coefficient of zero. Thus another example might be that you are predicting that the control group A has a greater mean value than group C; this specific hypothesis is that:

$$A > C$$

and hence (making the left-hand side greater than zero)

$$+1(A) + 0(B) - 1(C) > 0$$

The coefficients are labelled as λ_i for the i groups.

3. You can check that you have a valid set of coefficients, because if so, then they sum to zero, i.e. $\sum \lambda_i = 0$.

In addition, each contrast must be independent of all the others. This is the main reason why a maximum of $(i-1)$ contrasts are allowed, since it is not possible for more than that to be independent. Stat-istically, if two contrasts are independent they are called orthogonal. If you have equal sample sizes per group (as every well-designed experi-ment should have!), you can easily check whether any two contrasts (a and b) are independent since the following must be true:

$$\sum \lambda_{ia} \times \lambda_{ib} = 0$$

In the case of unequal sample sizes of n_i per group, the equivalent formula is $\sum n_i \times \lambda_{ia} \times \lambda_{ib}$, and there are additional complications later on in the calculation. Strictly it is not essential for every contrast to be independent, but if they are not, you need to adjust the threshold probability of significance: instead of 0.05, it becomes $1 - (0.95)^{1/r}$, where r = the number of non-independent contrasts you make.

4. Do a general parametric one-way ANOVA first (*see* Box 3.4a(i)), since technically you are breaking the among-group differences down into specific independent contrasts (*decomposing the sums of squares* in

statistical jargon). You will need the error mean square, MS_{error}, to test each contrast.

5. For each contrast, obtain the coefficients (λ_i) and using the mean values (m_i) for each of the original groups, do the following calculation, which results in a sum of squares for the contrast. This contrast has one degree of freedom, and hence is also a mean square (MS):

$$L = \Sigma \lambda_i \times m_i$$

If you are testing a *specific* hypothesis about the contrast, check at this point that L is positive. If it is, this means that the data follow the predicted pattern (analogous to the positive value of a one-tailed t-test). If it is not positive, then you know already that your hypothesis will not be supported by the data, and will not be significant.

$$MS_L = L^2 n / \Sigma \lambda_i^2$$

where n is the sample size of each group.

If you have unequal sample sizes for your groups, follow the method of Sokal and Rohlf (1995, pp. 528–9), but you should probably use a computer package to do this calculation for you!

6. Now form the variance ratio

$$F = MS_L / MS_{error}$$

where MS_L is the mean square for the contrast, and MS_{error} is the error mean square from the one-way ANOVA. The test statistic is then

$$t = \sqrt{F}$$

7. The degrees of freedom of the contrast is $a(n-1)$, where a is the number of groups involved in the contrast (excluding the groups left out, those with coefficients of zero).

   (a) If your hypothesis is *general*, then look up the critical value of a two-tailed t-test in Table D of Appendix III.

   (b) If your hypothesis is *specific*, then look up the critical value of a one-tailed t-test in Table D of Appendix III.

8. Repeat steps 2–7 for each contrast.

9. Present the results of the one-way ANOVA, as well as the contrasts, laid out in the standard manner (*see* worked example in Appendix II).

(i) Mean ranks: how to do a general nonparametric one-way ANOVA

Box 3.4b

1. Formulate the prediction. In this case, the general prediction being made is to ask whether there are any differences at all among the mean ranks of the groups. Therefore the null hypothesis actually tested is that there are no differences among the group mean ranks.

2. The test considers i groups of data, each of which contains n_i data values. The total number of data values in all the groups together = $N = \sum n_i$.

3. Rank all the values across all the groups combined (as in the U-test), giving low rank scores to low values. Once again, tied values are given the average of the ranks they would have been ascribed had they been slightly different. Where there are lots of tied values relative to the sample size, you may need to apply a tie-correction factor, but in this case it is better to get the calculation done by a computer.

4. Sum the ranks in each group, giving R_i in each case.

5. The test statistic is H, where

$$H = \frac{12}{N(N+1)} \times (\sum R_i^2 / n_i) - 3(N+1)$$

6. The degrees of freedom are $(i-1)$.

7. Look up the significance of the calculated H value as if it were a χ^2, in Table A of Appendix III (although H is not actually a χ^2, its value is distributed in the same way, so it is as if we were using χ^2). There is no standard layout for a general nonparametric ANOVA; just quote the test statistic, its degrees of freedom and the probability (*see* worked example in Appendix II).

8. Note that this is a *general* prediction and, therefore, if significant we can only conclude that there are *some* differences among the mean ranks: we are *not* allowed to say which particular pair of mean values are significantly different because we did not make any specific prediction beforehand about the ordering of the mean values, nor did we set up planned specific contrasts between particular sets of groups. Many researchers would like to be able to make such *post hoc* tests, however, and some methods are available (but not often used). They should only be used as a rough guide to what the differences might be, preferably to set up a priori hypotheses for a new data set to test. Sokal and Rohlf (1995, p. 431) and Day and Quinn (1989) have some recommendations.

(ii) Mean ranks: how to do a specific nonparametric one-way ANOVA

There are two ways of making specific predictions about the mean ranks of your groups. The one we favour here uses all the groups in a single a priori prediction of their rank order. The alternative is to use a priori contrasts (*see* point 9, below).

1. Formulate the specific prediction by specifying a particular rank order of the mean ranks of the groups, based on some a priori know-

ledge (theory, or previous published or gathered data), in advance of obtaining the data. The null hypothesis is that the rank order does not follow the prediction.

2. The test considers i groups of data, each of which contains n_i data values. There are N data values in total in all the groups (= Σn_i).

3. Rank all the values across all the groups combined (as in the U-test), giving low rank scores to low values. Once again, tied values are given the average of the ranks they would have been ascribed had they been slightly different. Where there are lots of tied values relative to the sample size, you may need to apply a tie-correction factor, but in this case it is better to get the calculation done by a computer.

4. Sum the ranks in each group, giving R_i in each case.

5. Assign the predicted rank order to the groups, from the lowest (rank = 1) to the highest (rank = i). This rank order then provides the λ_i coefficient values. Using these λ_i values, calculate:

the observed $L = \Sigma \lambda_i R_i$,

the expected $E = (N+1)(\Sigma n_i \lambda_i)/2$,

the variance $V = (N+1)(N \Sigma n_i \lambda_i^2 - (\Sigma n_i \lambda_i)^2)/12$.

6. Calculate the test statistic, z, as: $z = (L - E)/\sqrt{(V)}$. (Note that z is a standardized statistic which does not have any degrees of freedom.)

7. Look up the value of z in Table C of Appendix III, where you will see that the critical value for this specific (one-tailed) test is 1.64. If your value is greater than this, then the result is significant. If it is significant, we then reject the null hypothesis: there is evidence that the mean ranks fall into the predicted rank order.

8. What do you do if the result is *not* significant? If the mean ranks in fact fall in the opposite direction to your prediction, the value of z will be negative and quite possibly greater in absolute magnitude than 1.64. You *cannot* conclude *anything* about this, since your predicted rank order was not supported. You benefited from a gain in power over a general test, but the cost was that you could not conclude anything from a failure to reject the null hypothesis. You certainly cannot go and test an alternative rank order: this would now be *post hoc* since you have seen the actual pattern of the mean ranks. What you can do, following on from a non-significant specific test, is to ask the question: my predicted rank order was not supported, but is there evidence of *any* differences among groups in the data? In other words, you can go ahead and do a general test for any differences.

9. An alternative method is to use a priori contrasts, similar to the parametric case of Box 3.4a(ii). If done in advance of obtaining the data, these allow $(i-1)$ independent contrasts to be made, each one consisting of a subset of the groups contrasted against another subset.

Follow the method of Box 3.4a(ii), points 1–2, to obtain the co-efficients for the contrast you want to make, and then create two new groups by adding together the data for all the original groups that have the same sign (+ or –) coefficient. These two artificial groups are then tested using either a Mann–Whitney U-test (general prediction only), or preferably a nonparametric one-way ANOVA (general or specific prediction).

Note that the coefficients for these specific contrasts are *not* the same thing as the coefficients that specify the rank order for the test outlined above, points 1–8. You cannot use positive, negative and zero coefficients in the rank order test, imagining that you are doing a specific contrast.

10. There is no standard layout for a specific nonparametric one-way ANOVA; just quote the test statistic for each prediction, its degrees of freedom (if appropriate) and the probability. Note that a z-test does not have degrees of freedom, whereas specific contrasts have one degree of freedom (using either Mann–Whitney U-tests or nonpara-metric one-way ANOVA).

A worked example of a specific nonparametric one-way ANOVA is shown in Appendix II.

Post hoc *testing after analysis of variance.* A one-way analysis of variance will tell us whether or not there is a significant difference between our groups, but it does not tell us where the difference lies. Of course, inspection of the mean values for each group will give us a clue, but in many cases that is all it will give us. A glance at Fig. 2.4, for instance, tells us that marginal damage tends to increase with leaf size, but is the significant effect due to the damage in small leaves being disproportionately low, or to that in large leaves being dispro-portionately high, or what? It is possible to find out by conducting an appropriate *post hoc* comparison of each of the pairs of groups within the analysis. While this might sound suspiciously like the round robin comparisons we warned against above, the crucial difference is that *post hoc* comparisons are conducted within the overall analysis of variance and not as independent primary tests of significance. Where parametric analyses of variance are concerned, a plethora of potential *post hoc* tests are available. However, as with all statistical tests, they each have underlying assumptions and must be used with care and only as a guide to where the relevant differences might be. (Day and Quinn (1989), Sokal and Rohlf (1995: section 9.7) and Underwood (1997) provide very useful cautionary discussions.) Two of the most widely used are the Student Newman–Keuls test and the least signi-ficant difference test (*see* Box 3.4a), both of which are available, with other options, on many of the major statistical packages. Unfortunately, there are no generally accepted *post hoc* tests for nonparametric analyses of variance.

Covariates. Sometimes it is useful to control for other, nuisance, factors which are measured on a constant-interval scale within an analysis of variance. For instance, an analysis of differences in mating success between territory-owning and non-territory-owning male wood mice (*Apodemus sylvaticus*) might want to control for body size to rule out an effect of territory ownership arising simply because owners tend to be bigger. In a parametric analysis of variance, body size could be incorporated as a *covariate* (as long as it was normally distributed). The analysis would then reveal the independent effects of territory ownership and body size. Most major statistical packages allow covariates to be included in analyses of variance.

1 × n *chi-squared.* The one-way analysis of variance used a comparison of group means for the rank scores of individual data values to arrive at a test statistic. As in the two-group case, however, we could perform a chi-squared test on the *totals* for each group *as long as the data values are counts* (*see above*). We then have what is known as a 1 × n chi-squared analysis, where *n* is the number of groups. The two-group chi-squared earlier is just one form of this. The calculation of expected values and the test statistic χ^2 are exactly the same as for the two-group chi-squared test; thus we first decide whether expected values should be the average count in each case or whether there are reasons for having unequal expected values, then we calculate $\chi^2 = \Sigma(O - E)^2/E$ as before and check the outcome in a table (e.g. A in Appendix III) of χ^2 threshold values for $n - 1$ degrees of freedom.

Tests for differences in relation to two levels of grouping

In all the above difference tests, we were concerned with differences within a single level of grouping, e.g. between groupings based on seed colour. However many groups of seed colour we had (red, yellow, green, orange, blue, etc.) we would still be dealing only with seed colour and thus with one level of grouping. But we can easily envisage situations in which we would be interested in more than one level of grouping. For example, we might want to know not only whether chicks peck at some colours of grain more than others, but also whether pecking in males is different from that in females. More interestingly still, we might want to know whether the sex of chicks affects the *difference* in pecking at the various colours of grain. Is the difference stronger in one sex? Is it in the same direction in both sexes? Here we have *two* levels of groupings: seed colour and sex. In the examples that follow, we shall look at analyses that cater for two levels of grouping with two groups in each.

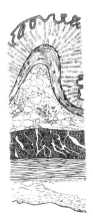

2 × 2 *chi-squared*. If we have data in the form of counts, we can again use chi-squared, but in a slightly different way from before. The table below shows the number of stomach tissue biopsies revealing cancerous cells in men and women who had been given one of two different chemotherapy treatments:

	Treatment A	Treatment B	Row totals
Men	20	17	37
	(exp. = 12.3)	(exp. = 24.7)	
Women	0	23	23
	(exp. = 7.7)	(exp. = 15.3)	
Column totals	20	40	60

Expected values in a 2 × 2 (or any $n \times n$) chi-square table are calculated in a rather different way from those in a $1 \times n$ chi-squared. First the rows and columns are totalled and the grand total calculated. The expected value for each cell in the table is then calculated as (row total × column total)/grand total. Thus for men given Treatment A, the expected number of biopsies revealing cancer is $(37 \times 20)/60 = 12.3$; for women given Treatment B it is $(23 \times 40)/60 = 15.3$. The test statistic is once again calculated as $\chi^2 = \Sigma(O - E)^2/E$, which in this case comes to 18.8. To arrive at the appropriate degrees of freedom, we multiply the number of rows minus one by the number of columns minus one. In a 2 × 2 table, the number of degrees of freedom is therefore $1 \times 1 = 1$; in a 3 × 4 table, it would be $2 \times 3 = 6$ and so on. Checking $\chi^2 = 18.8$ for 1 d.f. against Appendix III, Table A, shows that it is significant at the 0.1 per cent (0.001) level. We can thus conclude that men and women differ significantly in their incidence of cancer following the two treatments.

2 × 2 *two-way analysis of variance*. As before, if we want to compare sets of individual data values, we can use analysis of variance but this time it is a two-way rather than a one-way analysis. In a 2 × 2 two-way analysis of variance, the data are cast into four cells (two × two groups, which can be cast as two rows and two columns of a table). If we wanted to do a 3 × 5 two-way analysis the data would be cast into 15 cells, and so on for any combination of levels of grouping. Once again there are both parametric and nonparametric versions of the analysis (Box 3.5a, b). As usual, the parametric test assumes the data conform reasonably to normality. This assumption is, of course, relaxed for the nonparametric equivalent. However, *both* parametric and nonparametric tests assume that the data have the same variance (*see* Box 2.1) within each cell (i.e. within each combination of levels of grouping) – another example of the distribution-free, but not assumption-free, nature of nonparametric tests. Both types of analysis compare the mean values of the columns within the classification and the mean values of the rows. In other words they compare means within each of the two levels of grouping. Comparisons between the column means or between the row means are known as the *main effects* and are distinguished from a second kind of comparison referred to as an *interaction*. (Note that an interaction can be calculated only if all cells contain more than one datum value.) If there is a significant interaction it means the two sets of samples at one level of grouping respond differently to differences in the second level of grouping. An example makes the distinction between main effects and interaction clear. Imagine our two levels of grouping are freshwater versus marine fish and male versus female, and the variable for comparison is growth rate. Freshwater versus marine can be the rows of the classification (*see table below*) and male versus female the columns. The analysis would be concerned with the following: (a) main effect 1: differences in row means (is there any difference in growth rate between freshwater and marine fish?), (b) main effect 2: differences in column means (is there any difference in growth rate between males and females?) and (c) any interaction between the levels of grouping (e.g. is the difference in growth rate between males and females greater in one environment than in the other?).

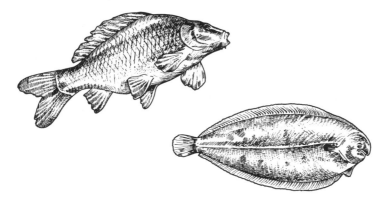

		Water type	
		Fresh	Marine
Sex	Male	a b c d mean = A	e f g h mean = B
	Female	i j k l mean = C	m n o p mean = D

As in the one-way analysis of variance, the nonparametric version can be used to test either general or specific predictions, but now about both the two main effects and any interactions. Making specific predictions is rather more involved in the two-way analysis because we need to be clear as to exactly what we are comparing and to calculate different coefficients for each specific comparison so predictions can be tested. Box 3.5b illustrates the procedure.

Although we have confined ourselves to a two-way analysis of variance here, the two-way model is only a particular case of a multifactor analysis of variance where there can be three, four or more levels of grouping. The principles underlying more complex analyses are the same, but as the number of levels of grouping increases it becomes more and more difficult to interpret the proliferating interactions terms. Moreover, the more levels of grouping that are included, the slimmer the chance they are all truly independent. Full discussion of these analyses can be found in Sokal and Rohlf (1995). In parametric analyses, as in the one-way case, it is also possible to incorporate covariates to control for factors on a constant interval scale.

Box 3.5a	(i) Mean values: how to do a general parametric two-way ANOVA

This is a little more complicated than a parametric one-way ANOVA, but in principle it is the same. Using the freshwater/marine fish example:

1. The two-way design is cast as ij cells (here 4) formed from i columns (here $i = 2$) and j rows (here $j = 2$) in a table. Each cell has a number of replicate measurements, n per cell.

   Formulate the predictions. In this case for example, (a) *rows*: marine fish differ in growth rate from freshwater fish; (b) *columns*: male fish differ in growth rate from female fish; (c) *interaction*: water type and sex of fish interact to determine growth rate. The appropriate null hypotheses are that there are no differences among the rows, or among the columns, and that there are no interactions of any kind.

2. Work out the mean values of all the replicates in each cell (B_{ij}), and the overall means of each column (C_i) and row (R_j) in the two-way design. The grand mean of all the data is M.

3. Calculate SS_{rows}, the sum of squares of the row mean values (not the raw data), following the method given in Box 3.1. Do the same for the column mean values to find SS_{cols}, and the cell means to find SS_{grps}.

4. Calculate the interaction sum of squares, $S_{int} = SS_{grps} - SS_{rows} - SS_{cols}$.

5. Calculate the sum of squares of each cell separately, and add them up to give SS_{error}.

6. The degrees of freedom for columns $\text{d.f.}_{\text{cols}} = (i - 1)$, for rows $\text{d.f.}_{\text{rows}} = (j - 1)$, for the interaction $\text{d.f.}_{\text{int}} = (i - 1)(j - 1)$. The error degrees of freedom $\text{d.f.}_{\text{error}} = ij(n - 1)$.

7. Calculate the mean squares: $MS_{\text{rows}} = SS_{\text{rows}}/\text{d.f.}_{\text{rows}}$; $MS_{\text{cols}} = SS_{\text{cols}}/\text{d.f.}_{\text{cols}}$; $MS_{\text{int}} = SS_{\text{int}}/\text{d.f.}_{\text{int}}$; $MS_{\text{error}} = SS_{\text{error}}/\text{d.f.}_{\text{error}}$.

8. Calculate the test statistic for each component:

    $F_{\text{rows}} = MS_{\text{rows}}/MS_{\text{error}}$ with degrees of freedom of $\text{d.f.}_{\text{rows}}$, $\text{d.f.}_{\text{error}}$

    $F_{\text{cols}} = MS_{\text{cols}}/MS_{\text{error}}$ with degrees of freedom of $\text{d.f.}_{\text{cols}}$, $\text{d.f.}_{\text{error}}$

    $F_{\text{int}} = M_{\text{int}}/MS_{\text{error}}$ with degrees of freedom of d.f._{int}, $\text{d.f.}_{\text{error}}$

9. Look up these values of F with the appropriate degrees of freedom in Table G of Appendix III to see whether they exceed the relevant critical values. Present the two-way ANOVA laid out in the standard manner (*see* worked example in Appendix II).

10. Note that if the interaction is significant, then the row and the column effects don't really mean much, because the difference among the row groups will then depend on which column group it is, and vice versa.

11. You should be aware that there are two kinds of grouping factors that create the 'ways' of a one- or two-way ANOVA. The distinction doesn't make a difference in a one-way parametric analysis, but it does in a two-way. First, there are *fixed* factors, where the groups are fixed by the experiment, and are not intended to represent a few of a great range of possibilities; the hypothesis being tested is about whether real differences exist among the actual groups in the experiment. Many experimentally created groupings are of this type, such as a treated and a control group (which are not representative of a range of different possible groups). A *random* factor, on the other hand, has groups that are merely a sample of all possible groups, and are intended to represent this range of possibilities; here one is not really interested in whether there are differences among these particular groups, but rather in how much of the variation is contained among as opposed to within the groupings. Rearing sets of animals in different cages would be a good example: cage differences are not really treatments, but merely random variation that is not interesting but nevertheless must be allowed for. Sokal and Rohlf (1995, Section 8.4) and Underwood (1997) have good discussions of this distinction.

(ii) Mean values: how to do a specific parametric two-way ANOVA

As in the one-way case (Box 3.4a(ii)), we cannot specify a particular rank order that we expect, but we can make a priori contrasts. If these contrasts are specified in advance of obtaining the data, they allow $(i - 1)$ contrasts to be made among column groups, $(j-1)$ contrasts among row groups, and $(i-1)(j-1)$ contrasts involving the interaction of both row and column groups. As before, each contrast consists of a subset of the groups contrasted against another subset.

1. Formulate the predictions that you want to make in the form of contrasts, before collecting the data. You do this by expressing the prediction as an inequality, just as in Box 3.5b(ii). From rearranging the inequality, you obtain the relevant set of coefficients, λ_i, for each contrast. Unlike Box 3.5b(ii), however, you can either cast each contrast as a general or a specific contrast (using $\neq$ or $<$, respectively, in the inequality, just as in Box 3.4a(ii), point 2).

2. Each contrast must be valid and independent of all the others (using the checks exactly as detailed in Box 3.4b(ii), point 3).

3. Do a general parametric two-way ANOVA first (*see* Box 3.5a(i)), since you are technically breaking the among-group differences down into specific independent contrasts ('decomposing the sums of squares' in statistical jargon). You will need the MS_{error} term from this analysis.

4. For each contrast, obtain the coefficients (λ_i) and the mean values (m_i) for each group, and follow exactly the method detailed for the one-way case in Box 3.4a(ii), points 5 and 6, to obtain the test statistic, t, and its significance for your contrast.

5. Repeat step 4 for each contrast.

6. Present the results of the two-way ANOVA, as well as the contrasts, laid out in the standard manner (*see* worked example in Appendix II).

Box 3.5b

(i) Mean ranks: how to do a general nonparametric two-way ANOVA

1. Rank the data values in all cells combined and add up the ranks in each cell to give a rank total R for that cell.

2. We can now use these rank totals to test our *general* predictions:

   (a) *Marine fish differ from freshwater fish.*
   Sum the rank totals for each *column* (*see table above*) separately giving an R_i value for marine and an R_i value for freshwater environments. Now calculate H as in the one-way analysis of variance (Box 3.4b), using the column R_i values and their appropriate n_i values (here there are eight values making up each R_i, hence $n_i = 8$), and check the resulting value in an appropriate table (e.g. Appendix III, Table A) as if it were χ^2 for $(i-1)$ degrees of freedom.

   (b) *Male fish differ from female fish.*
   Sum the rank totals for each *row* separately giving an R_i value for male and an R_i value for female fish and calculate H again as above. Again, $n_i = 8$ in our example.

(c) *There is an interactive effect of water type and sex on the growth rate of fish.*

This is an open-ended general prediction which combines both (a) and (b) above. It is asking whether there is any interaction between levels of grouping in determining growth rate. The calculation of H is exactly as above but the $\sum R_i^2/n_i$ term includes the rank totals for *all* the cells in the table instead of just the columns or the rows to give H_{tot}. Here n_i is again the number of values making up each R_i ($= 4$ in our example). H for the interaction, H_{int}, can then be calculated as: $H_{int} = H_{tot} - H_{water} - H_{sex}$, where H_{water} and H_{sex} refer to H values from (a) and (b) above. The degrees of freedom for H_{int} are $\text{d.f.}_{tot} - \text{d.f.}_{water} - \text{d.f.}_{sex}$, in this case, therefore, $3 - 1 - 1 = 1$ (d.f._{tot} is the total i (number of group means) minus 1).

(ii) Mean ranks: how to do a specific nonparametric two-way ANOVA

Testing our *specific* predictions is a little more complicated but still relatively straightforward. We give three illustrations below:

(a) A prediction about the columns, e.g. *marine fish grow faster than freshwater fish* (or vice versa). In terms of the means in the table above this predicts that $(B + D) > (A + C)$. Another possibility is the opposite prediction, that freshwater fish grow faster than marine, i.e. $(A + C) > (B + D)$.

(b) A prediction about the rows, e.g. *male fish grow faster than female fish* (or vice versa). This predicts $(A + B) > (C + D)$ (or, conversely, for the prediction that females grow faster than males, $(C + D) > (A + B)$).

(c) A prediction about interaction, e.g. *the effect of water type on growth rate will be greater in male fish than in female fish.* This predicts that $(A - B) > (C - D)$. The converse (the effect will be greater in females) would, of course, predict $(C - D) > (A - B)$. This class of prediction is thus concerned with the *interaction* between water type and sex.

These are the predictions we can make about the relative sizes of the means; how do we arrive at the coefficients for testing? The procedure is as follows:

1. The first step is to rearrange the various predicted inequalities so that all the means are on the left, thus:

    (a) The prediction about the effect of water type becomes $-A + B - C + D > 0$ (for growth in marine > growth in fresh water) or $+A - B + C - D > 0$ (for growth in fresh water > growth in marine).

    We then substitute 1 with the appropriate sign for each letter so that we arrive at $-1, +1, -1, +1$ or $+1, -1, +1, -1$, respectively.

(b) In the same way, the prediction about the effect of sex becomes $+A + B - C - D > 0$ (for males > females) or $-A - B + C + D > 0$ (for females > males) and the coefficients thus $+1, +1, -1, -1$ or $-1, -1, +1, +1$.

(c) The interaction predictions must also be framed in this way. Thus the prediction $(A - B) > (C - D)$ becomes $+A - B - C + D > 0$ and the coefficients $+1, -1, -1, +1$. The prediction $(C - D) > (A - B)$ becomes $-A + B + C - D$ and the coefficients therefore $-1, +1, +1, -1$.

2. Then, testing these predictions

(a) *Marine fish grow faster than freshwater fish.*
Remember this means we are testing the prediction $-A + B - C + D > 0$. The coefficients λ_i thus become $-1, +1, -1, +1$ so that the rank totals for each cell are weighted as follows:

$$L = \sum \lambda_i R_i$$

$$= (-1)(R_{\text{freshwater/male}}) + (+1)(R_{\text{marine/male}})$$

$$+ (-1)(R_{\text{freshwater/female}}) + (+1)(R_{\text{marine/female}})$$

E and V can then be calculated as:

$$E = (N + 1)(\sum n_i \lambda_i)/2$$

$$V = (N + 1)[N\sum n_i \lambda_i^2 - (\sum n_i \lambda_i)^2]/12$$

The test statistic z can then be calculated as:

$$z = (L - E)/\sqrt{V}$$

and checked against a table of z-values (*see* Appendix III, Table C).

(b) *Male fish grow faster than female fish.*
Now we are testing the prediction $+A + B - C - D > 0$, so λ_i becomes $+1, +1, -1, -1$. The calculation of L, E and V and then the test statistic z can proceed as above, but with the new λ_i weightings.

(c) *The effect of water type is greater in males than in females.*
This tests the interaction prediction $+A - B - C + D > 0$ using λ_i of $+1, -1, -1, +1$. Once again, follow the calculations above for L, E, V and z.

There is no standard layout for a nonparametric two-way analysis of variance; just quote the test statistic, its degrees of freedom and the probability.

A full worked example of a 2×2 two-way analysis of variance can be found in Appendix II.

3.3.4   Tests for a trend

As with analysis of differences, there are many tests that cater for trends. We shall introduce two simple ones here, both looking at the relationship between two sets of data. More complex versions of these tests allow multiple relationships to be tested at the same time, but they are beyond the scope of this book.

Correlation analysis

The first test is one of correlation. Correlation analyses calculate a test statistic known as a correlation coefficient, the two most commonly used being the parametric Pearson's product-moment correlation coefficient r and the nonparametric Spearman rank correlation coefficient r_s (Box 3.6). A correlation coefficient quantifies the extent to which there is an association between two sets of data values. A large *positive* coefficient indicates a strong tendency for high values in one set to co-occur with high values in the other and low values in one set to co-occur with low values in the other. A large *negative* coefficient indicates a strong tendency for high values in one set to co-occur with low values in the other and vice versa. Correlation coefficients take a value between +1.0 and −1.0, with values of +1 and −1 indicating respectively a perfect positive or negative association. 'Perfect association' means that every value in one set is predicted perfectly by values in the other set. A coefficient of 0 indicates there is no association between the two sets of values so that values in one set cannot be predicted by those in the other. If a correlation is significant, it implies that the size of the coefficient differs significantly (positively or negatively) from zero, the value expected under the null hypothesis.

As a parametric test, the Pearson product-moment correlation is subject to a number of assumptions about the data being analysed. The two sets of data must be normally distributed individually *and* jointly (a bivariate normal distribution in the jargon), both must be measured on a constant interval scale and the relationship between them must be linear. Satisfying these assumptions can be a tall order and the Pearson correlation is probably used more liberally than it should be. Being a nonparametric test, the Spearman rank correlation can be used with ordinal (ranking) or constant interval measurements and, of course, is not sensitive to departures from normality. Importantly, the relationship also need not be linear, but merely continuously increasing or decreasing (monotonic); thus r_s will test for many sorts of curved patterns of association.

| Box 3.6 | **How to calculate a correlation coefficient** |

Set out the two sets of data values to be correlated in pairs (remember, for each value in set 1 there must be a corresponding value in set 2). Thus, if we were looking for a correlation between height and weight in people, the data would be set out as below:

Person	Weight (kg)	Height (m)
1	63	1.8 (pair 1)
2	74	1.7 (pair 2)
3	60	1.9 (pair 3)
4	71	1.8 (pair 4)
.	.	. .
.	.	. .
.	.	. .
		etc.

It may be that several values of one measure are paired with the same value of the other, for example when measuring some behaviour in several individuals from the same social group and using these values in a correlation of time spent doing the bahaviour and group size. In this case, the data might be as follows:

Observation	Time spent in behaviour (s)	Group size
1	15.3	3 (pair 1)
2	17.1	3 (pair 2)
3	18.0	5 (pair 3)
4	6.0	5 (pair 4)
5	31.1	5 (pair 5)
.	.	. .
.	.	. .
.	.	. .
		etc.

We can now calculate either a parametric (Pearson) or nonparametric (Spearman rank) correlation:

(i) Pearson correlation coefficient (parametric)

This tests for a *linear* association between two (bivariate-) *normally* distributed variables.

1. Formulate the prediction, either as a general (any) or a specific (positive or negative) association. You can test for either a general

(two-tailed) or specific (one-tailed) correlation coefficient by using different threshold values of the test statistic.

2. Calculate S_{xx}, S_{yy} and S_{xy} as shown in Box 3.2.

3. Calculate the test statistic, $r = S_{xy}/\sqrt{(S_{xx} \times S_{yy})}$

4. Look up the threshold value for r using $(n-2)$ degrees of freedom in Table E of Appendix III, using either the one-tailed or two-tailed levels of significance, as appropriate to your hypothesis (*see* worked example in Appendix II).

(ii) Spearman rank correlation coefficient (nonparametric)

This tests for a *monotonic* (i.e. continuously increasing or decreasing) association between two variables, and works with ranking or constant-interval measurements, making no assumptions about the normality of the data. As with the Pearson coefficient, we can test for either a general (two-tailed) or specific (one-tailed) correlation by using different threshold values of the test statistic.

1. Rank the values for the first measure only, then rank the values for the second measure only.

2. Subtract second-measure ranks from first-measure ranks (giving d_i) then square the resulting differences (d_i^2) and calculate the Spearman coefficient as:

$$r_S = 1 - [(6\Sigma_i d_i^2)/(n^3 - n)]$$

where n is the number of pairs of data values.

3. If n is between 4 and 20 (4 is a minimum requirement), consult Appendix III, Table F for the appropriate sample size to see whether the calculated r_S value is significant. Remember that general (two-tailed) or specific (one-tailed) tests have different threshold values for r_S. Thus you must know whether you are looking for any association at all (general), or just a positive, or just a negative one (specific).

4. If n is greater than 20, a different test statistic, t, is calculated from r_S as:

$$t = r_S\sqrt{[(n-2)(1-r_S^2)]}$$

t can then be checked against its own threshold values (Appendix III, Table D) for $n-2$ degrees of freedom.

A significant calculated value for r_S or, with large samples, t, allows us to reject the null hypothesis of no trend in the relationship between our two measures. A fully worked example of a Spearman rank correlation analysis is given in Appendix II.

While we can test for trends with correlation analyses, we must interpret them with care. Two things in particular should always be borne in mind. First, a correlation does not imply cause and effect. While we may have reasons for *supposing* that one measure influences the other rather than vice versa (*see* earlier discussion of *x* and *y* measures in trend analysis), a significant correlation cannot be used to confirm this. All it can do is demonstrate that two measures are associated. A well-known example illustrates the point. Suppose we acquired some data on the number of pairs of storks breeding in Denmark each year since 1900, and also the number of children born per family in Denmark in the same years. Plotting a scattergram, with breeding storks as the *x*-axis and babies as the *y*-axis, and calculating a correlation coefficient reveals a significant positive correlation at $p < 0.001$ between the two measures. Do we conclude that storks bring babies? Of course not! All we can conclude is that, over the period examined, there is some association between the number of breeding storks and the human birth rate; perhaps both species simply reproduce more during long, hot summers! Second, as we have said, correlation analyses assume that associations between measures are linear (or reasonably so) if the test is parametric, or at least monotonic if the test is nonparametric. If they are neither, a lack of significant correlation cannot be taken to imply a lack of association. This is made clear in Fig. 3.4 in which the relationship between two measures is U-shaped. A correlation coefficient for this would be close to zero, but this doesn't mean there is no association. The book by Martin and Bateson (1993) contains a very useful discussion of these and other problems concerning correlation analyses.

Linear regression

Correlation analyses allow us to judge whether two measures are associated, but that's all. Two important things that they do not allow us to establish are: (a) whether changes in the value of one measure *cause* changes in the value of the other and (b) anything quantitative about the association.

Cause and effect. In the case of the storks and babies above, it is fairly clear that changes in *x* (the number of breeding storks) do not cause changes in *y* (the number of human babies): the association arises either through some indirect cause-and-effect relationship (e.g. hot, sunny years encourage both storks and humans to breed) or through trivial coincidence. In many cases, though, it is not so clear whether there is or is not a direct cause-and-effect relationship between *x* and *y*. The best way to decide is to do an experiment where, instead of merely measuring *x* and *y* as in correlation analysis, we experimentally change *x*-values and measure what subsequently

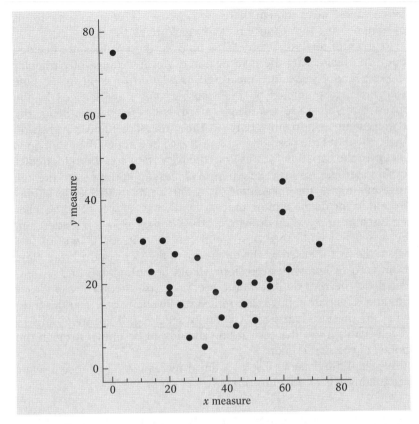

Figure 3.4 Associations between two variables need not be linear. Correlation analysis would not reveal a significant trend, but an association between the variables appears to exist (*see text*).

happens to y. If we see that y changes when x is changed, we can be reasonably happy with a cause-and-effect interpretation. This is quite different from correlation analysis.

Quantitative relationships.   In addition, correlation analysis does not allow us to say much about the quantitative relationship between x and y (if x changes by n units, by how much does y change?) or allow us to predict values not included in the original trend analysis (e.g. can we predict the response of an insect pest to a 40 per cent concentration of pesticide when an analysis of the effect of pesticide concentration goes up to only 30 per cent?). With certain qualifications, linear regression analysis may allow us to do all the above. The qualifications arise mainly from the usual assumptions made by parametric tests, though the requirement for normality applies to one of the data sets only (the y-axis measure). However, like correlation analysis, and as its name implies, linear regression also requires the relationship to

be passably linear, though there are ways of overcoming some forms of nonlinearity, for instance by log-transforming the data.

Regression analysis proceeds as follows: the x-values are decided upon in advance by the investigator to cover a range of particular interest, and y-values are measured in relation to them to see how well they are predicted by x. Because x-values are selected by the investigator, they are regarded as fixed, error-free values, hence the requirement of normality only on the y measure. Linear regression then calculates the position of a line of best fit through the data points and uses the equation for this line (the *regression equation*) to predict other values for testing. The criterion of 'best fit' in this case is the line that minimizes the magnitude of positive and negative deviations from it in the data – a more precise, mathematical way of doing what we attempt to do when drawing a straight line through a scattergram by eye. The significance of a trend can be assessed in one of two ways: the first is based on the difference in the *slope* of the best fit line from zero (a line with zero slope would be horizontal) and is indicated by the test statistic t; the second tests whether a significant amount of variation in y is accounted for by changes in x, and is indicated by the test statistic F. Generally speaking, if the trend fails to reach significance, the best fit line should not be drawn through the points of the scattergram.

We shall use F as our test for a linear regression. The procedure for calculating F is outlined in Box 3.7.

Box 3.7

(i) How to do a parametric linear regression

1. Calculate S_{xx}, S_{yy} and S_{xy} as in Box 3.2.
2. Calculate the slope of the line as: $b = S_{xy}/S_{xx}$.
3. Calculate the intercept of the line on the y-axis as: $a = \bar{y} - b\bar{x}$, where $\bar{y}$ is the mean of the y-values and $\bar{x}$ is the mean of the x-values.

The line can now be fitted by calculating $y = a + bx$ for some sample x-values and drawing it on the scattergram.

To calculate the standard error of the slope:

4. Calculate the *variance of* y *for any given value of* x as:
$s^2_{y/x} = [1/(n-2)][S_{yy} - S^2_{xy}/S_{xx}]$.
5. The *standard error of the slope* is then: $\sqrt{(s^2_{y/x}/S_{xx})}$.
6. To find the test statistic F, calculate the following:

Regression sum of squares (RSS) $= (S_{xy})^2/S_{xx}$

Deviation sum of squares (DSS) $= S_{yy} - (S_{xy})^2/S_{xx}$

Regression mean square (RMS) $=$ RSS$/1$

(1 is the value always taken by the *regression degrees of freedom*.)

Deviation mean square (DMS) = DSS/$(n - 2)$.

($n - 2$ is the value taken by the *deviation degrees of freedom*.)

F is now calculated simply as F = RMS/DMS and its value can be checked against critical values in F-tables (Appendix III, Table G) for 1 (f_1) and $n - 2$ (f_2) degrees of freedom.

7. To find y for new values of x:

    Having established our regression equation, we might well want to predict y for other values of x that lie within the range we actually used in the analysis. Once we had then gone away and *measured* y for our new x-value we should want to see whether it departed significantly from its predicted value. Three steps are needed:

    (a) calculate the predicted y-value using the equation $y = a + bx$ as when fitting the regression line, but this time use the new x-value (x') in which you are interested;

    (b) calculate the standard error (s.e.) of the predicted y-value as follows:

    $$\text{s.e.} = \sqrt{[(s_{y/x}^2)(1 + 1/n + d^2/S_{xx})]}$$

    where $d = x' - \bar{x}$;

    (c) calculate the test statistic t as:

    $$t = \frac{\text{observed } y - \text{predicted } y}{\text{s.e.}}$$

    and look up the calculated value of t in t-tables (Appendix III, Table D) for $n - 2$ degrees of freedom (where n is the number of pairs of data values in the regression). If t is significant it means the measured value of y departs significantly from the value predicted by the regression equation and might lead to interesting questions as to why.

A worked example of a regression analysis can be found in Appendix II.

(ii) How to do a nonparametric regression

A nonparametric approach to regression is little used (*see* Sokal and Rohlf, 1995, p. 539). There seem to be few advantages over using a simple rank correlation in cases where you do not want to predict a value of y from x, or to discover the actual equation of their relationship. However, some use is made of a nonparametric procedure (spline regression) fitted to a set of points to produce a smoothed description of highly nonlinear irregular relationships (*see* Schluter, 1988 and Pentecost, 1999).

3.4 Testing hypotheses

The previous section has introduced an armoury of basic significance tests with which we can undertake confirmatory analyses of differences and trends. Knowing that such tests are available, however, is not much use unless we know how and when to employ them and gear our data collection to meet their requirements. It is important to stress again, therefore, that the desired test(s) should be borne in mind from the outset when experiments and observations are being designed and the data to be collected decided upon.

3.4.1 Deciding what to do

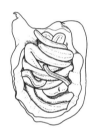

Having arrived at some predictions from our hypotheses, we must decide how best to test them. This sounds straightforward in principle but involves making a lot of careful decisions. Are we looking for a difference or a trend? What are we going to measure? How are we going to measure it? How many replicates do we need? What do we need to control for? There is no general solution to any of these problems; the right decision depends entirely on the prediction in hand and the material available to test it. In a moment, we shall go back to the predictions we derived from our observational notes to see how we can test some of them. Before doing that, however, we should be aware of some important principles and pitfalls of experimental/observational design and analysis.

Significance, sample sizes and statistical power

As should be obvious from what we've said already, much hinges on the quality of the data sample we have at our disposal. It should be as representative of the population from which it derives as possible if we're to stand a chance of coming to sensible conclusions. But what does that mean? How many data values make a representative sample? What sample size do we need to make our tests for differences or trends reasonably powerful, i.e. actually capable of detecting the effects we're testing for? Sadly, there is usually no simple answer. Vague rules of thumb, such as 'at least 20 values per sample', can be and have been suggested (e.g. Dytham, 1999), but they are just that, vague rules of thumb.

The whole picture of what significance is, and what you the experimenter can do about it, is summarized in the table below.

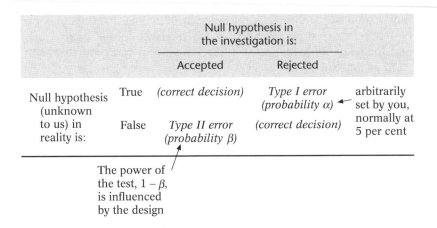

		Null hypothesis in the investigation is:		
		Accepted	Rejected	
Null hypothesis (unknown to us) in reality is:	True	*(correct decision)*	*Type I error (probability α)*	arbitrarily set by you, normally at 5 per cent
	False	*Type II error (probability β)*	*(correct decision)*	

The power of the test, $1 - \beta$, is influenced by the design

The design of the experiment tries to match up these situations so that correct decisions are made as many times as possible. The chance (α) of rejecting the null hypothesis when it is really true (making a Type I error) is the 5 per cent criterion we discussed above (*see What is statistical significance?*). As we mentioned there, this probability is arbitrarily set at 5 per cent in most cases. Making it smaller would on the face of it lead to fewer Type I errors, but would then also mean that the opposite mistake is made more often, accepting the null hypothesis when really the hypothesis is false (a Type II error, with a chance of occurring of β).

Luckily, Type II errors are also influenced by the design of the experiment. The *power* of the design to detect real differences, $1 - \beta$, depends on the sort of statistical test you are using (different kinds of test have very different power), the variation within your sample (the greater it is, the lower the power), the effect size you are trying to detect (the larger it is, the greater the power) and the sample size (the larger it is, the greater the power). You can't do much about the effect size or the variance except measure it accurately, but you can choose the most powerful test, and usually you have control over the number of data values that you collect.

If you have preliminary estimates of the effect size and the variation, you can work out the sample size that will result in a given level of power – 80 per cent power to detect the effect you are looking for is regarded as a minimum level that your experiments should try to achieve. There are some good programs that can help do these calculations, either within a statistical package (e.g. StatGraphics, SPSS) or as a dedicated program (e.g. nQuery). It is important to make the effort to conduct power analyses. Many otherwise admirable studies have been compromised by hopelessly small sample sizes for the variability of the data. Just as unnecessarily, effort and experimental subjects have been wasted through data overkill. Further useful discussion of power analyses and their importance in designing investigations can be found in Sokal and Rohlf (1995), Underwood (1997) and Tomas and Juanes (1996) (the last also discuss various power analysis software packages).

Some dangerous traps

Confounding effects.   One of the commonest problems in collecting and analysing data is avoiding so-called confounding effects. Confounding effects arise when a factor of interest in an investigation is closely correlated with some other factor which may not be of interest. If such a correlation is not controlled for, either in the initial design of an investigation or by using suitable techniques during analysis, the results will inevitably be equivocal and any potential conclusions compromised. For example, suppose we wanted to know whether the burden of a particular parasitic nematode increased with the body size of the host (e.g. a mouse). We might be tempted simply to assay worm burden and measure body size for a number of arbitrarily chosen host individuals, and then perform a correlation analysis. If we got a significant positive correlation coefficient, we might conclude that burden increased with body size. An unwelcome possibility, however, is that host body size correlates with age so that bigger hosts also tend to be older. If there is some age-related change in immune competence (e.g. older hosts are less able to resist infection), a positive correlation with size could arise that actually has nothing to do with host body size. In this case size is confounded with age and simple correlation analysis cannot disengage the two. The best solution here would be to select different sized hosts from a given age group so that the confounding effect is controlled for from the outset. Order effects are common confounding factors in many undergraduate projects. Testing animals in, say, treatment 1 first, then in treatment 2, then in 3, etc., confounds treatment with time. Animals may simply be tired by the time they get to treatment 3, so any difference in their response to the treatment could be due to that.

Floor and ceiling effects.   Floor and ceiling effects arise when observational or experimental procedures are either too exacting or too undemanding in some way to allow a desired discrimination to be made. For example, looking for differences in mathematical ability among people by asking them the solution to $5 + 3$ is unlikely to be very fruitful because the problem is too easy; everyone will get the right answer straightaway. A ceiling effect (everyone performs to a high standard) will thus prevent any differences there may be in ability becoming apparent. Conversely, if the same people were asked to solve a problem in catastrophe theory the odds are that no one would be able to do it. In this case, a floor effect (everyone does badly) is likely to prevent discrimination of ability. Floor and ceiling effects are not limited to performance-related tasks. Similar limitations could arise in, for instance, histological staining. If a particular tissue requires just the right amount of staining to become discriminable from other tissues, the application of too little stain would result in everything appearing similarly pale (a floor effect) while too much stain would result in everything appearing similarly dark (a ceiling effect). Real differences in tissue type would thus not show up at the extremes of stain application. Floor and ceiling effects are clearly a hazard to be avoided and are well worth testing for in preliminary investigations.

Non-independence.   One of the commonest sources of error in data collection and analysis arises from non-independence of data values. In many circumstances, there is a temptation to treat repeated measures taken from the same subject material as independent values during statistical analysis. As Martin and Bateson (1993) point out, this error arises from the misconception that the aim of a scientific observation or experiment is to obtain large numbers of measurements rather than measurements from a large number of subjects. The point is, of course, that obtaining additional measures from the same subject is not the same as increasing the number of subjects in the sample. An example of such an error would be as follows. Suppose an investigator wished to assess the average rate of nutrient flow in the phloem of a particular plant species. Setting up a preparation might be involved and time-consuming. To save effort, the investigator decides to take as many measurements as possible from each preparation before discarding it. As a result, there are 15 measurements from one preparation, 10 from another and 12, 16 and 5 from three more. To calculate the average, the investigator totals the measurements and divides by $n = 58$ (i.e. $15 + 10 + 12 + 16 + 5$). Of course, the measurements from each preparation are not independent; there may be something about the plant in each case that gives it an unusually high or low rate of nutrient flow relative to most of the plants in the population. Incorporating each measurement taken from it as an independent example of flow rate in the population as a whole is clearly going to bias the average upwards or downwards. The true n-size in the above example is five (the number of preparations) not 58. Measurements from each preparation should thus be averaged, or collapsed in some other way, to provide a single value for use in analysis. The fallacy of this kind of approach becomes obvious if we consider estimates of average plant height rather than nutrient flow rate. Few people would seriously measure the height of the same plant 16 times and regard these as independent samples of the height of the

species concerned. The principle, however, is exactly the same in the flow rate example and is referred to as *pseudoreplication*.

The problem with non-independence is that it can operate at several different levels and sneak insidiously into analyses unless careful attempts are made to exclude it. We have discussed it only at the level of the individual subject. However, depending on what is being measured, it could arise if, for example, related individuals are used as independent subjects or if plants grown on the same seed tray or animals kept in the same cage are used. We should thus be on our guard against it at all times.

3.5   Testing predictions

Having highlighted some potential pitfalls, we must now bear them in mind as we return to our main observational examples and design experiments to test some of their predictions. We shall take one prediction from each.

Example 1

Plants and herbivores

E.g. Prediction 1A(ii)   *Leaf damage by slugs will decrease the further up a plant that samples are taken.*

This predicts a negative trend between leaf damage and height up the plant. The assumption is that the height of a leaf off the ground influences its vulnerability to slugs. Two approaches immediately suggest themselves. We could conduct a survey in the field, measuring the height of leaves above the ground and scoring the amount of slug damage on each, or we could carry out an experiment, in the field or the laboratory, exposing leaves at different heights to slugs in a controlled environment. Either way, there is a formidable number of factors to take into account if we're to get a sensible outcome.

The most obvious is that our exploratory samples came from a wide range of plant species. At the very least there are likely to be confounding effects of species-specific attributes such as the presence of distasteful toxins or other deterrents. A first consideration in a field survey, therefore, might be to select plants of a similar range of heights within each of several species. This would ensure that the confounding effects of height and species were removed, but on its own it

would still not be enough for a robust comparison. A major uncontrolled factor remains: the prevalence of slugs. Different species of plant are likely to occupy different habitats, some of which are more suitable for slugs than others. Another potential confounding effect may therefore need to be removed. To check, we could carry out a simple census of slug populations around our subject plants. If numbers did not differ significantly, we could happily ignore them. If they did differ, however, we should need some way of taking them into account. One way would be to weight the recorded damage by the observed prevalence of slugs before analysis. We could analyse the relationship between weighted damage and height as a trend, using a correlation or regression analysis, but an alternative approach, that would allow us to look in more detail at the effect of plant species, would be to use two-way analysis of variance. A two-way analysis of variance, with height of leaves above the ground (low, medium, high) and plant species as the two levels of grouping, would reveal the separate main effects of height and plant species, but also the (very likely) interaction between them. The interaction term would probably be very important here because structural and developmental differences between species will almost certainly influence the effects of position up the plant. An alternative way to take slug prevalence into account in a parametric analysis of variance would be to include it as a covariate, a constant interval measurement whose effect on the data can be controlled for within the analysis of main effects and interaction.

If height did emerge as a significant predictor of slug damage, it would, of course, lead to further questions. Is the height effect due to slugs being unwilling to climb beyond a certain height? Is it due to leaves further up being tougher or more noxious? Is it due to taller plants having greater gaps between successive leaves so discouraging further ascent by slugs? Any height × species interaction might offer a clue to some of these (and other) possibilities by highlighting species characteristics that increase or decrease the height effect. An easier way to get at them, however, might be to do some laboratory experiments.

A laboratory study would attempt to control things more tightly at the outset. One approach might be to cultivate individual plants of some of the species sampled so that they were of similar height and the important morphological characteristics, such as the number and spacing of leaves, were, as far as possible with different species, standardized. Plants of different species could then be arranged randomly or in a regular, alternating pattern on a bench, so that any systematic confounding of position and species was avoided, and each plant was exposed to the same number of similarly sized slugs for a set period, say overnight, and then scored for leaf damage. We might be tempted simply to catch a few slugs in the field and use those. However, this would be unwise. Freshly caught slugs would enter the experiment with an unknown feeding history. We would know neither their level of hunger, nor what they had recently been feeding on. Both factors could introduce unwelcome bias into the experiment or even cause a floor effect. The best thing to do would be to bring slugs into the laboratory well before the experiment (or culture them in the laboratory), feed them all on the same material (a combination of the plant species to be tested) and deprive them of food for a short time (e.g. 12 hours) prior to testing. All slugs would then be standardized for feeding experience and hunger.

The design above would allow us to assess the effect of leaf height on damage and, if we chose to observe slug activity (directly or using, for example, time-lapse photography), we might be able to conclude something about how any effect came about. Suppose we found that higher leaves did indeed sustain less damage, but that this was due not to slugs failing to get up to them but to slugs feeding for a shorter time when they did get there. Two possible explanations might be: (a) slugs were nearly satiated by the time they reached the higher leaves or (b) higher leaves are less palatable. An easy way to test for the latter would be to present slugs with standard-sized discs of material cut from leaves at different heights. We could choose three heights (high, medium and low) on different, but standard-sized, plants. If higher leaves are less palatable, discs cut from them would sustain less damage within the test time.

Example 2

Hosts and parasites

E.g. Prediction 2A *Parasite burdens will increase with host testosterone levels.*

This prediction derives from the hypothesis that reproductive hormones might influence susceptibility to infection. Both sex and stress hormones are known to affect the immune system, often in concert, though their effects on resistance to parasites are very variable. Prediction 2A is based on the observation that adult voles in the samples generally had greater parasite burdens than juveniles, and males greater burdens than females (Fig. 2.1). Since the difference between age classes is much more pronounced in males, testosterone becomes a plausible candidate for driving the effects of age and sex on parasite burdens. As with the seemingly simple prediction about leaf height and herbivore damage, however, much needs to be thought about in testing whether there is a connection.

We could start by taking some animals from the field and assaying their parasite burdens and testosterone levels. This could be done nondestructively by taking blood samples and faeces for blood and gut parasites, and inspecting the fur for ticks, fleas and other ectoparasites. Circulating testosterone concentrations could be assayed from

either the blood or the faecal samples. Since testosterone secretion is highly pulsatile, and we are interested in chronic effects, faecal samples might provide the more appropriate measure since they accumulate testosterone metabolites over a period. Depending on the degree of discontinuity of testosterone concentration across age and sex classes, and the extent to which it can be normalized, we could test for testosterone as a predictor of age and sex differences in parasite burden by including it as a covariate in a two-way analysis of variance. The variable of interest would be our measure of parasite burden and the levels of grouping sex and age class. If we first ran the analysis without testosterone and found significant age and sex effects (as Fig. 2.1 suggests we might), but then found that the effects disappeared and were replaced by a significant covariate effect of testosterone when the latter was included, we should have some evidence that testosterone was important in generating our initial age and sex differences in parasite burden. If the distribution of testosterone values did not permit this, we could instead test for differences in testosterone between age and sex classes using nonparametric analysis of variance and seek correlations between testosterone levels and parasite burden within classes.

Such analyses might tell us something, but it would be limited. One of the main problems is that simply taking animals from the field and testing for associations between hormone levels and parasites does not allow us to say anything about cause and effect. Testosterone may well influence parasite burden, but equally both may correlate with something else, such as level of social activity, which affects exposure to infection. The association would then be an artefact of testosterone levels and parasite burden being linked to exposure. A further problem is that testosterone may covary with circulating levels of corticosterone, a glucocorticoid 'stress' hormone that also influences the immune system. The best way to control for these potentially confounding effects is either to manipulate levels of testosterone and corticosterone directly, by injection or slow-release implants, and then see what happens when animals are given a controlled infection with a known parasite, or to monitor spontaneous levels of the hormones over a period and challenge in the same way. In order to use either approach, however, animals would need to be cleaned of existing parasites and given a period of acclimation before any experiment. Corticosterone plays a role in the immune response to many infections, particularly gut helminths (worms), and any residual infection is likely to compromise investigations of hormonal effects on resistance. The best approach to start with would probably be to monitor spontaneous levels of testosterone followed by challenge. While manipulating levels experimentally allows selective control of individual hormones, it can also cause unwanted side effects and disrupt the delicate interactions between physiological systems that underpin hormonal effects on resistance. It may also be necessary to remove relevant endocrine tissue surgically or by chemical ablation to prevent spontaneous secretion affecting the control of circulating levels. Such manipulations might thus be better as a follow-up to test conclusions arising from the more observational approach. The assumption in the latter, of course, is that differences in spontaneous circulating levels of hormone will predispose individuals to correspondingly different degrees of resistance. We should then look for an association between hormone levels prior to infection and the subsequent severity of the infection. Regression analysis would be the obvious candidate, perhaps using a suitable multivariate model to take the effects of both testosterone and corticosterone levels into account simultaneously.

Example 3

Nematodes and pollutants

E.g. Prediction 3B *Species present at unpolluted sites but missing from polluted sites will show greater mortality when exposed to pollutants.*

This prediction assumes that pollution is causally responsible for the absence of certain species from polluted samples. While this seems simple enough to test, we might want to explore its basis a little more before embarking on a set of experiments.

The most obvious point is that, with samples from only three sites, pollution status and site are confounded. Particular species may be present or absent at a given site purely by chance, or because sites happened to differ in some other important respect (e.g. interstitial water content of the soil) that affected their viability for different species. Ideally, therefore, we should first replicate our samples within site categories by choosing a number of sites, say four to six, of each kind (unpolluted, polluted with heavy metals, polluted with organophosphate) across which there is some variation in other environmental features. Species that are consistently absent from polluted sites in these samples would provide a better basis for further investigation.

One way forward might then be to collect or culture representative nematode species that are present only at unpolluted sites and expose

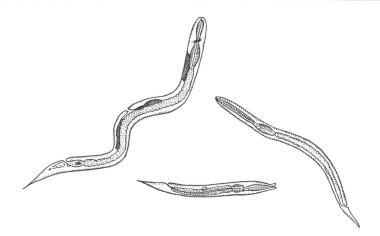

them, say in standardized Petri dish cultures, to representative concentrations of heavy metal or organophosphate pollutant, not forgetting a suitable control (e.g. distilled water). Each treatment might be replicated half a dozen times. One-way analysis of variance of mortality by treatment would then reveal any significant effect due to the experimental pollutants. Significantly greater mortality in the two polluted treatments would be evidence in favour of a direct impact of pollution on species survival and thus presence/absence at particular sites. However, a lack of any effect would not necessarily rule out pollution as being responsible for the absence of certain species from polluted sites. Simply bathing adult worms in solutions and seeing if they die is a crude approach to say the least. Pollutants may work at any of a number of points in the worms' life cycle, perhaps reducing fecundity (the number of eggs produced) or the survival of larvae. Similar experiments could be performed to test these possibilities. More subtly, the effects may depend on particular environmental conditions, for example interactions between pollutants and other chemicals in the soil. This may be the case even though the simple experiment above showed a mortality effect; bathing worms in raw pollutant may kill them, but this may not be the way they are killed by pollution in the field. More complex experimental treatments, simulating patterns of exposure in the soil, might thus be called for.

E.g. Prediction 4C *Encounters will progress further when opponents are more similar in size and it is more difficult to judge which will win.*

Example 4
Crickets

This prediction derived from observing that encounters between male crickets followed an apparently escalating pattern from chirping

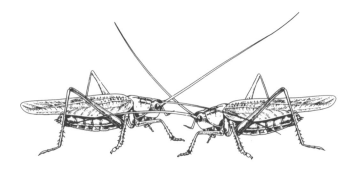

and antenna-tapping to out-and-out fighting and that, on the whole, bigger crickets tended to win. A possibility, therefore, is that progressive escalation reflects information-gathering about the relative size of an opponent and the likelihood of winning if the encounter is continued. If relative size is difficult to judge, as when two opponents are closely matched, the likelihood of winning cannot be judged in advance and the only way to decide the outcome is to fight. The prediction is thus of a negative trend between degree of escalation and the relative size of opponents: degree of escalation should increase with decreasing difference in size.

At first sight, this seems easy enough to test using a Spearman rank correlation or regression analysis. However, we first need some way of measuring degree of escalation. So far all we have are behavioural descriptions – chirping, antennating, fighting, etc. – from which we have inferred levels of escalation. Somehow we must put numbers to these. It is clear that we cannot put the behaviours on some common constant interval scale; we cannot, for instance, say that antennating is twice as escalated as chirping and fighting ten times as escalated. The easiest thing is simply to rank them. Thus what we assume to be the lowest level of escalation, say chirping, takes a rank of 1 and the highest level a rank of n, where n is the number of levels we decide to identify. We can then use the ranks of 1 to n as the y-values in our trend analysis.

To obtain our x-values, we must decide on a suitable measure of size. Ideally, the measure should be reliable and repeatable within and between individuals; measuring the size of the flexible abdomen, for instance, might not be a good idea because this could vary with food and water intake and thus vary from one encounter to another. It would be better to measure some component of the hard exoskeleton, e.g. the length of the long hind leg or the width of the thorax, which will not vary over the time course of observations. Of course, in our analysis, we are interested in a measure of the *relative* size of opponents, so our x-values must be some measure of relative size. The most obvious might be the *difference* in size between opponents. However, it is not hard to see why this would be inadequate. Suppose we observed two crickets of thorax widths 7.5 and 8.5 mm respectively. Suppose we observed another pair of thorax widths 6.5 and 5.5 mm. In both cases, opponents differ by 1 mm and would score the same on a simple difference measure. In the first case, however, 1 mm is only 6 per cent of the combined width measures; in the second it is 8 per cent. A 1 mm difference may thus create a greater asymmetry in the likelihood of winning in the second case than in the first. As a result it would be better to use a ratio rather than a difference scale on the x-axis, e.g. size of bigger opponent/size of smaller opponent.

Having decided on our measures, we can now plan observations. Since we are looking for a trend, we want to end up with pairs of x- and y-values. One way we might proceed is to put a number of individually marked males into a sand-filled arena and record all encounters over, say, 20 minutes, noting the males involved and the highest level (on our scale of 1 to n) to which each encounter progressed. One problem with this approach, however, is that some males would interact more than once. The pairs of x- and y-values arising from each repeat encounter could not be used independently because body size ratio would be confounded with pair of opponents and escalation levels might be influenced by the males' past experience of each other. It would therefore be better to arrange encounters between different pairs of males to provide independent replicates of a range of size ratios. Since we are using ordinal (rank) measures of y and selected ratios as x, we should test for significance in our trend using a Spearman rank correlation rather than regression.

3.6   Refining hypotheses and predictions

The discussions above do two things. First they give some simple indications as to how to set about testing particular predictions. Second, they show that the outcome of such tests differ in the extent to which they increase or decrease our confidence in the hypothesis from which the prediction derives. Thus, for example, a failure to find an association between leaf damage and height up a plant in a field survey (Prediction 1A(ii)) would not greatly undermine our confidence in the hypothesis (1A) that leaf damage reflected availability to slugs. A host of factors besides height up the plant is likely to affect attack by slugs. Prediction 1A(ii) is thus a very restrictive test of Hypothesis 1A. The important point, however, is that the process of testing Hypothesis 1A does not stop there. It isn't abandoned just because one rather simplistic prediction did not work out. Instead the predictions are refined, gradually ruling out confounding factors. We saw

this in the plants and herbivores example, as the suggested investigations used laboratory experiments to test for effects of height by controlling for changes in the size, spacing and palatability of leaves further up the plant.

Refinement also takes place in the opposite direction. A prediction that *is* borne out does not necessarily offer direct support for its parent hypothesis. The test of Prediction 3B (nematodes and pollutants) is a good example. This predicts that the species of nematode absent from polluted sites will die when exposed to pollutants in the laboratory. It derives from the general hypothesis (3A) that pollution reduces species diversity. The finding that pollutants kill these species when they are bathed in them in Petri dishes does not necessarily mean that direct susceptibility to pollutants is the reason for reduced species diversity at polluted sites, or even the absence of those particular species from such sites. Their susceptibility in the laboratory may reflect a more general susceptibility to stressors. A variety of chemicals, quite unrelated to the environmental pollutants in question, might have a similar effect and this should certainly be tested. The absence of a particular species from polluted sites may reflect an interaction between effects of pollutants and competitive ability, with absence ultimately being due to competitive exclusion by other species rather than mortality through pollution. A more complex experimental design comparing effects on putatively robust (present at polluted sites) as well as putatively susceptible species would increase confidence in a mortality effect if differential mortality of susceptible, but not robust, species emerged.

Although we have presented hypotheses and their predictions in a rather cut-and-dried fashion through this book, it is clear that there is really considerable fluidity in both. The relationship between hypothesis, prediction and test is a dynamic one and it is through the modifying effects of each on the others that science proceeds.

3.7 Summary

1. Predictions derived from hypotheses dictate the experiments and observations needed to test hypotheses. As a result of testing, hypotheses may be rejected, provisionally accepted or modified to generate further testable hypotheses.

2. Decisions about experimental/observational measurement and the confirmatory analysis of such measurements are interdependent. The intended analysis determines very largely what should be measured and how, and should thus be clear from the outset of an investigation.

3. Some kind of yardstick is needed in confirmatory analysis to allow us to decide whether there is a convincing difference or trend in our measurements (i.e. whether we can reject the null hypothesis of no difference or trend). The arbitrary, but generally accepted, yardstick is that of statistical significance. Significance tests allow us to determine the probability that a difference or trend as extreme as the one we have obtained could have occurred purely by chance. If this probability is less than an arbitrarily chosen threshold, usually 5 per cent but sometimes 1 or 10 per cent, the difference or trend is regarded as significant and the null hypothesis is rejected.

4. Different significance tests may demand different attributes of the data. Parametric and nonparametric tests differ in the assumptions they make about the distribution of data values within samples and the kinds of measurement they can cope with. Tests can also be used in specific/one-tailed or general/two-tailed forms depending on prior expectations about the direction of differences or trends.

5. Basic tests for a difference include chi-squared, the t-test, the Mann–Whitney U-test and one- and two-way analysis of variance.

Each has a number of requirements which must be taken into account and care is needed not to make multiple use of two-group difference tests in comparing more than two groups of data.

6. Basic tests for a trend include Pearson product-moment and Spearman rank correlations and regression analysis. Correlation is used when we merely wish to test for an association between two variables. Regression is used when we change the values of one variable experimentally and observe the effect on another to test for a cause and effect relationship between two variables. Regression analysis yields more quantitative information about trends than correlation but makes more stringent demands on the data.

7. Testing predictions requires careful thought about methods of measurement, experimental/observational procedure, replication and controlling for confounding factors and floor and ceiling effects.

8. Whether or not a test supports or fails to support a prediction, the implications for the parent hypothesis need careful consideration. A hypothesis does not necessarily stand or fall on the outcome of one prediction; everything depends on how discriminating the prediction really is.

References

Bailey, N. T. J. (1981) *Statistical methods in biology*, 2nd edition. Hodder and Stoughton, London.

Day, R. W. and Quinn, G. P. (1989) Comparisons of treatments after an analysis of variance in ecology. *Ecological Monographs* **59**, 433–463.

Dytham, C. (1999) *Choosing and using statistics: a biologist's guide.* Blackwell, Oxford.

Gaines, S. D. and Rice, W. R. (1990) Analysis of biological data when there are ordered expectations. *American Naturalist* **135**, 310–317.

Martin, P. and Bateson, P. (1993) *Measuring behaviour*, 2nd edition. Cambridge University Press, Cambridge.

Meddis, R. (1984) *Statistics using ranks: a unified approach.* Blackwell, Oxford.

Pentecost, A. (1999) *Analysing environmental data.* Longman, Harlow.

Schluter, D. (1988) Estimating the form of natural selection on a quantitative trait. *Evolution* **42**, 849–861.

Siegel, S. and Castellan, N. J. (1988) *Nonparametric statistics for the behavioral sciences*. McGraw-Hill, New York.

Sokal, R. R. and Rohlf, F. J. (1995) *Biometry*, 3rd edition. Freeman, San Francisco.

Tomas, L. and Juanes, F. (1996) The importance of statistical power analysis: an example from animal behaviour. *Animal Behaviour* **52**, 856–859.

Underwood, A. J. (1997) *Experiments in ecology*. Cambridge University Press, Cambridge.

CHAPTER 4

Presenting information

How to communicate outcomes and conclusions

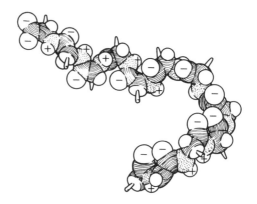

If your investigation has gone to plan (and possibly even if it hasn't), you will have generated a pile of data which somehow needs to be presented as a critical test of your hypothesis. Performing appropriate significance tests is only one step on the way. While significance tests will help you decide whether a difference or trend is interesting, this information still has to be put across so that other people can evaluate it for themselves. There are two reasons why we should take care how we present our results. The first is to ensure we get our message over; there is little point making a startling discovery if we can't communicate it to anyone. The purpose of our investigation was to test a hypothesis. We might conclude that the results support the hypothesis or that they undermine it. Whichever conclusion we reach we must sell it if we wish it to be taken seriously. Since scientists are by training sceptical, selling our conclusion may demand some skilful presentation and marshalling of arguments. The second reason is that we must give other people a fair chance to judge our conclusions. As we saw earlier, simply saying that some difference or trend is significant doesn't tell us how strong the effect is. It is important to present results in such a way that others can make up their own minds about how well they support our conclusions. In this chapter, we shall look at some conventions in presenting information that help satisfy both these requirements. We begin with some simple points about figures and tables.

4.1   Presenting figures and tables

We stressed earlier that it is usually not helpful to present raw data. Raw data are often too numerous and the information in them too difficult to assimilate for useful presentation. Instead, we summarize them in some way and present the summary form. We have already dealt with summary statistics in a general way when we discussed exploratory analysis; here we discuss their use in presenting the results of confirmatory analysis.

Although it is usually obvious that raw data require summarizing, there can be a temptation to summarize everything, as if summary statistics or plots were of value in themselves. In confirmatory analyses, they are of value only to the extent that they help us evaluate tests of hypotheses. We thus need to be selective in distilling our results. Naturally, the summaries and forms of presentation that are most appropriate will depend on the type of confirmatory analysis. It is therefore easiest to deal with different cases in turn.

4.1.1   Presenting analyses of difference

Where we are dealing with analyses of difference, the important information to get across is a summary of the group values being compared. The form this takes will vary with the number of groups and levels of grouping involved. There are two basic ways of presenting a summary of differences: figures and tables. As we have argued previously, figures tend to be easier to assimilate than tables, even when the latter comprise summary statistics. However, tables may be more economical when large numbers of comparisons are required, or where comparisons are subsidiary to the main point being argued but helpful to have at hand. If tables are used, it is important that they present all the key summary information necessary to judge the claims they make. This usually means (a) summary statistics (e.g. means ± standard errors*, medians ± confidence limits) for each of the groups

---

*   While there are various forms of summary statistic, means ± standard errors are widely used because the mean of a set of values is an easy concept to grasp and because the standard error estimates the distribution of *means* within the population which approaches a normal distribution as the sample size increases. It is thus usually legitimate to quote means ± standard errors as summary statistics even when the distribution of data *values* demands a nonparametric significance test.

being compared, (b) the sample size (n) for each group, (c) test statistic values, (d) the probabilities (p-values) associated with the test statistic values and (e) an explanatory legend detailing what the table tells us. The test statistics and p-values can be presented either in the table itself or in the legend. The same information, of course, should be presented in figures except that the summary statistics are represented graphically (e.g. as bar charts) instead of as numbers, and information about sample sizes, test statistics and probability levels more conventionally goes in the legend (now usually called the figure caption) rather than in the figure itself. (Nevertheless, as long as it doesn't clutter the figure and detract from its impact, it can be very helpful to include statistical information within the figure and we shall do this later where appropriate.)

Differences between two or more groups (with one level of grouping)

Here we are presenting the kinds of result that might emerge from a Mann–Whitney U-test, or a one-way analysis of variance. Suppose we have tested for a difference in growth rate (general prediction) between two groups of plants, one given a gibberellin (growth-promoting) hormone treatment, the other acting as an untreated control. The treated group contained 12 plants and the untreated group eight. Using a one-way analysis of variance for two groups, we discover a significant difference at the 0.1 per cent ($p < 0.001$) level between the groups, with treated plants growing to a mean ($\pm$ standard error) height of 14.75 ± 0.88 cm during the experimental period, and controls growing to a mean height of 9.01 ± 0.63 cm. We could present these results as in Table 4.1a. Note the legend explaining exactly what is in the table.

Table 4.1a The mean height to which plants grew during the experimental period when treated with gibberellin or left untreated

| | Experimental groups | | |
	Treated	Untreated	Significance
Mean ($\pm$ s.e.) height (cm)	14.75 ± 0.88	9.01 ± 0.63	$H = 12.22,$
n	12	8	$p < 0.001$

The significance column could be omitted from the table, in which case the test statistic and probability level should be given in the legend. The legend would now read:

> Table 4.1a The mean height to which plants grew during the experimental period when treated with gibberellin or left untreated. H comparing the two groups $= 12.22$, $p < 0.001$

An alternative and frequently adopted convention in presenting significance levels is to use asterisks instead of the test statistic and probability numbers. In this case, different levels of probability are indicated by different numbers of asterisks. Usually * denotes $p < 0.05$, ** $p < 0.01$ and *** $p < 0.001$, but this can vary between investigations so it is important to declare your convention when you first use it. Using the asterisks convention, the table would now read as in Table 4.1b. Exactly the same forms of presentation, of course, could be used for comparisons of more than two groups. However, if we had

Table 4.1b The mean height to which plants grew during the experimental period when treated with gibberellin or left untreated. ***, $H = 12.22$, $p < 0.001$

	Experimental groups		
	Treated	Untreated	Significance
Mean (± s.e.) height (cm)	14.75 ± 0.88	9.01 ± 0.63	***
n	12	8	

Table 4.2a The number of plants surviving treatment with different herbicides

	Experimental group				
	Herbicide 1	Herbicide 2	Herbicide 3	Control	Significance
Number of plants surviving	15	8	6	27	$\chi^2 = 19.3$, $p < 0.001$

Table 4.2b The number of plants surviving treatment with different herbicides. ***, $\chi^2 = 19.3$, $p < 0.001$

	Experimental group				
	Herbicide 1	Herbicide 2	Herbicide 3	Control	Significance
Number of plants surviving	15	8	6	27	***

used a U-test to test for a difference between two groups, we should now have to change to a one-way analysis of variance to avoid abuse of a two-group difference test (*see* Chapter 3).

The presentation of means and standard errors (or medians and confidence limits) is appropriate whenever we are dealing with analyses that take account of the variability within data samples. In chi-squared analyses, however, where we are comparing simple counts, there is obviously no variability to represent. If we were presenting a chi-squared analysis of the number of plants surviving each of three different herbicide treatments and one control treatment, therefore, the table would be as shown in Table 4.2a. Table 4.2b shows an alternative presentation. If we wish to present our results as figures rather than tables, we can convey the same information using simple bar charts. Thus Table 4.1a can be recast as Fig. 4.1a. Similarly, Table 4.1b could be recast as Fig. 4.1b. For a comparison of three groups, say comparing the effectiveness of the lambda bacteriophage in killing three strains of *Escherichia coli* suspected of differing in susceptibility, the figure might be as shown in Fig. 4.2a.

In Figs 4.1a, b and 4.2a, we have assumed a one-way analysis of variance was used to test a general prediction (hence the test statistic H). If we had instead tested a specific prediction because we had an a priori reason for expecting a rank order of effect (e.g. gibberellin-treated plants would grow taller than untreated plants (Fig. 4.1), or

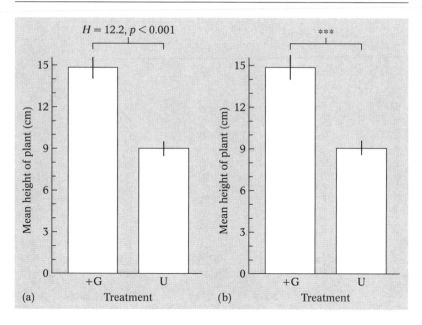

Figure 4.1 (a, b) The mean height to which plants grew during the experimental period when treated with gibberellin (+G, $n = 12$) or left untreated (U, $n = 8$). ***, $H = 12.2$, $p < 0.001$. Bars represent standard errors.

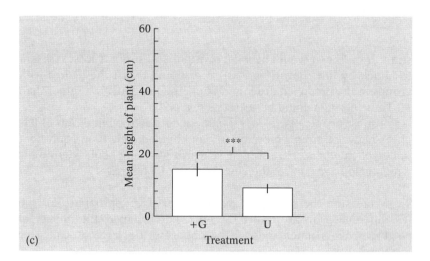

Figure 4.1 (c) Figure 4.1b with a different scale. The mean height to which plants grew during the experimental period when treated with gibberellin (+G, $n = 12$) or left untreated (U, $n = 8$). ***, $H = 12.2$, $p < 0.001$. Bars represent standard errors.

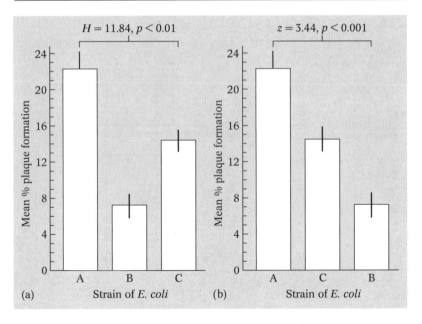

Figure 4.2   The mean percentage area of plaque (= bacterial death) formation by lambda bacteriophage on three strains (A–C) of *E. coli*. Bars represent standard errors. (a) Nonparametric analysis of variance testing a general prediction of difference between strains. $N = 8$ cultures in each case. (b) Nonparametric analysis of variance testing a specific prediction of difference between strains (A > C > B). $N = 8$ cultures in each case.

strain B of *E. coli* would be most resistant and strain A least resistant to attack by the phage (Fig. 4.2)), we should recast the figures with groups in the predicted order and quote the test statistic z rather than H. Thus Fig. 4.2a could be recast as Fig. 4.2b.

For most people, bar charts like these convey the important differences between groups more clearly and immediately than equivalent tables of numbers. However, it is worth stressing some key points which help to maximize the effectiveness of a figure.

1. Make sure the scaling of numerical axes is appropriate for the difference you are trying to show. For instance, the impact of Fig. 4.1b is much reduced by choosing too large a scale (*see* Fig. 4.1c).

2. Always use the *same* scaling on figures that are to be compared with one another. Thus, Fig. 4.3a, b is misleading because the different scaling makes the magnitude of the bars look the same in (a) and (b). Using the same scale, as in Fig. 4.3c, shows that there is in fact a big difference between (a) and (b).

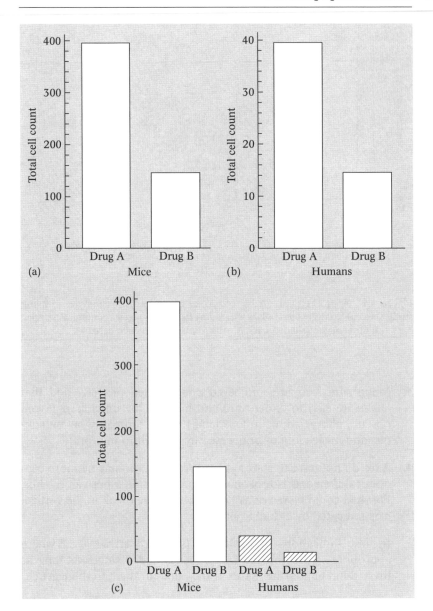

Figure 4.3   (a) The total number of T-helper cells in experimental samples from laboratory mice following administration of two different cytotoxic drugs (A and B). $N = 40$ samples for each drug treatment. (b) As Fig. 4.3a, but for experimental samples from humans. (c) The total number of T-helper cells in experimental samples from laboratory mice (open bars) and humans (shaded bars) following administration of two different cytotoxic drugs (A and B). $N = 40$ samples for each drug treatment and species.

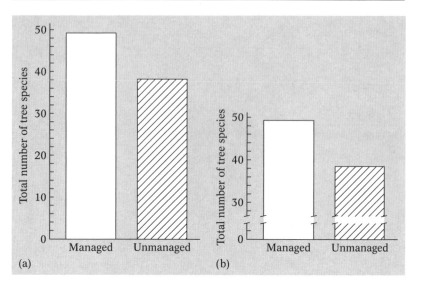

Figure 4.4   (a, b) The total number of species of tree bearing epiphytes in 2 km²
study areas of rainforest in Bolivia where forests are managed economically (open
bar) and unmanaged (shaded bar). $N = 1 \times 2$ km² area in each case.

3. Make sure axes are numbered and labelled properly and that
    labels are easy to understand and indicate the units used. Avoid
    obscure abbreviations in axis labels: these can easily be ambigu-
    ous and misleading or unnecessarily difficult to interpret.

4. Axes do not have to start at zero. Presentation may be more eco-
    nomical if an axis is broken and starts at some other value. Thus,
    Fig. 4.4a could be recast as Fig. 4.4b with the break in the vertical
    axis indicated by a double slash.

5. Include indications of variability (standard errors, etc.) where
    appropriate. Also include sample sizes and p-values as long as
    these don't clutter the figure. If they do, put them in the legend.

6. Always provide a full, explanatory legend. The phrasing of the
    legend should be based on the prediction being tested and the
    legend should include any statistical information (sample sizes,
    test statistics, etc.) not included in the figure. Do not repeat in-
    formation in both figure and legend, though. The legend should
    allow a reader to assess the information in the figure without hav-
    ing to plough through accompanying text to find more detailed
    discussion.

Differences between two or more groups (with more than one level of grouping)

Here we are concerned with the sort of results that might arise from a two-way analysis of variance or $n \times n$ chi-squared analysis. Presentation is now a little trickier because of the number of comparisons we need to take into account. A table is probably the simplest solution. For instance, suppose we had carried out a two-way analysis of variance looking at the difference in the frequency of accidental egg damage between three strains of battery hen maintained in three different housing conditions. Here we have two levels of grouping (strain and housing condition) with three groups at each level. The analysis tests for a difference between strains (controlling for housing condition), a difference between housing conditions (controlling for strain) and any interaction between the two levels of grouping (*see* Chapter 3). The best way to present the differences between groups within levels is to tabulate the summary statistics for each of the nine (3×3 groups) cells and include the test statistics in the legend. Thus testing for any difference between groups (i.e. not predicting a difference in any particular direction) might give the results shown in Table 4.3.

Table 4.3 The mean ($\pm$ s.e.) percentage number of eggs broken during the experimental period by three strains of battery hen (1–3) under three different housing conditions (A–C). Parametric two-way analysis of variance shows a significant effect of both strain ($F = 145.09$, d.f. $= 2,27$, $p < 0.001$) and housing ($F = 103.29$, d.f. $= 2,27$, $p < 0.001$) and a significant interaction between the two ($F = 58.76$, d.f. $= 4,27$, $p < 0.001$). $N = 4$ in each combination of strain and housing condition

		Strain		
		1	2	3
	A	43.50 ± 2.32	1.25 ± 0.75	22.25 ± 2.14
Housing condition	B	38.75 ± 1.09	13.75 ± 1.38	16.50 ± 1.71
	C	6.25 ± 0.85	10.25 ± 1.80	6.25 ± 1.80

This analysis reveals significant effects of both strain and housing conditions on egg breakage. These are obvious from the summary statistics in the table: breakage in strains 1 and 3 is relatively high under housing conditions A and B but drops sharply in condition C. In contrast, breakage in strain 2 is highest in conditions B and C and lowest in A. Damage tends to be greater in types A and B housing than in type C. In addition to these main effects, however, there is also a significant interaction between strain and housing condition (*see* legend to Table 4.3) with the effect of housing differing between strains. Although interaction effects can also be gleaned from a table of summary statistics like Table 4.3, they can be presented more effectively as a figure; one of the levels of grouping constitutes the *x*-axis and the measure being analysed is the *y*-axis. The relationship between the measure and the *x*-axis grouping can then be plotted for each group in the second level. Figure 4.5 shows such a plot for the interaction in Table 4.3. The lines in the figure, of course, simply indicate the groups of data: they are in no way comparable with statistically fitted lines. Full details of the analysis are given in the legend

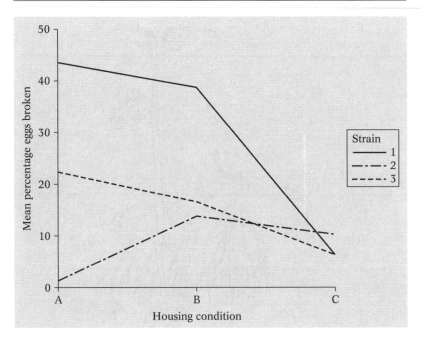

Figure 4.5 The mean percentage number of eggs broken by three strains of battery hen (solid, strain 1; dash/dot, strain 2; dotted, strain 3) in three different housing conditions (A–C). Parametric two-way analysis of variance showed a significant effect of both strain ($F = 145.09$, d.f. $= 2,27$, $p < 0.001$) and housing condition ($F = 103.29$, d.f. $= 2,27$, $p < 0.001$) and a significant interaction between the two ($F = 58.76$, d.f. $= 4,27$, $p < 0.001$). $N = 4$ in each combination of strain and housing condition.

because such a figure would not normally be presented as well as the summary table since it repeats information already given in the table. From Fig. 4.5 it is clear that, while all three strains show differences in egg damage across housing conditions, the direction and degree of decline are different in different strains. This implies that the effect of housing condition varies with strain, which is what is meant by an interaction between housing condition and strain.

With an $n \times n$ chi-squared analysis, we are just dealing with total counts in each cell so there are no summary statistics to calculate and present. The simplest presentation is thus an $n \times n$ table with each cell containing the observed and expected values for the particular combination of groups (the expected value in each cell usually goes in brackets). Table 4.4 shows such a presentation for a chi-squared analysis of the effects of temperature and soil type on the number of seeds out of 150 germinating in a seed tray.

Table 4.4   The number of seeds germinating in a tray in relation to temperature (low, 5 °C; high, 25 °C) and soil type. Expected values in brackets. $\chi^2 = 14.38$, d.f. = 1, $p < 0.001$

	Number of seeds germinating	
	In clay soil	In sandy soil
Low temperature	40	100
	(57.97)	(82.03)
High temperature	131	142
	(113.03)	(159.97)

4.1.2   Presenting analyses of trends

Presenting trend analyses is rather simpler because, in most cases, a scattergram with or without a fitted line is the obvious format. When it comes to more complicated trend analyses that deal with lots of different measures at the same time, summary tables rather than figures may be necessary. However, these need not concern us here.

Presenting a correlation analysis

Since correlation analysis does not fit a line to data points, presentation consists simply of a scattergram, though depending on how we have replicated observations this may include some summary statistics (*see below*). Information about test statistics, sample sizes and significance could be given in the figure, but it is more usual to include it in the legend. Thus Fig. 4.6a shows a plot of the number of food items obtained by male house sparrows (*Passer domesticus*) in relation to their dominance ranking with other males in captive flocks of six (rank 1 is the most dominant male that tends to beat all the others in aggressive disputes and rank 6 is the least dominant that usually loses against everyone else). In this case, observations were repeated for three sets of males so there are three separate points (*y*-values) for each *x*-value in the figure.

Although there is a significant trend towards dominant males getting more food, the correlation is negative because we chose to use a rank of 1 for the most dominant male and a rank of 6 for the least dominant. Rankings are frequently ordered in this way, leading to the slightly odd situation of concluding a positive trend (e.g. dominants get more food) from what looks like a negative trend (the number of food items decreases with increasing rank number). There is no reason, of course, why dominance shouldn't be ranked the other way round (6 = most dominant, 1 = least dominant) so that a positive slope actually appears in the figure.

Sometimes when replicated observations are presented in a scattergram, they are presented as a single mean or median with appropriate standard error or confidence limit bars. Thus an alternative presentation of Fig. 4.6a is shown in Fig. 4.6b. Note that a different explanatory legend is now required because the figure contains different information.

In some cases, replication may not occur throughout the data set. Say we decided to sample a population of minnows (*Phoxinus phoxinus*) in a stream to see whether big fish tended to have more parasites. To avoid the difficulties of making accurate measurements of fish size in the field and possibly injuring the fish, we visually assess those we catch as belonging to one of six size classes. We then count the signs of parasitism on them and return them to the water. Because we have no control over the number of each size class we catch, we end up with more samples for some classes than for others. When we come to present the data, we could present them as individual data points for each size class (Fig. 4.7a) or condense replicated data for classes to means or medians (Fig. 4.7b). In the latter case, only some points might have error or confidence limit bars attached to them because only some classes are replicated (*see* Fig. 4.7a). Data for those classes that are not replicated are still presented as single points.

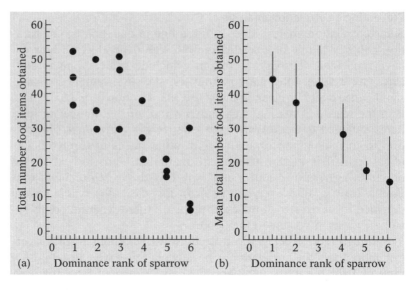

Figure 4.6   (a) The number of food items obtained during the period of observation by male house sparrows of different dominance status in groups of six (rank 1, most dominant; rank 6, least dominant, data for three groups at each rank). $r_s = -0.77$, $n = 18$, $p < 0.001$. (b) The mean number of food items obtained during the period of observation by male house sparrows of different dominance status in groups of six (rank 1, most dominant; rank 6, least dominant). $r_s = -0.77$, $n = 18$, $p < 0.001$. Bars represent standard errors.

It is, of course, important to remember that even where correlations are presented as mean or median values rather than independent data points, the correlation analysis itself (i.e. the calculation of the correlation coefficient) is still performed on the independent data points, not on the means or medians. Values of n are thus the same in Figs 4.6a and 4.6b and in Figs 4.7a and 4.7b. Correlations *can* be performed on summary statistic values, but obviously a lot of information is lost from the data and n-sizes are correspondingly smaller.

Presenting a regression analysis
Presenting a regression analysis is essentially similar to presenting a correlation except that a line needs to be fitted through the data points. If the trend isn't significant, so that a line should not be fitted, a figure probably isn't necessary in the first place. The details of

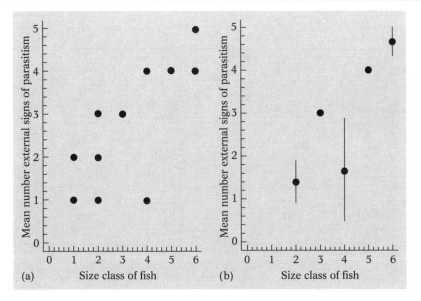

Figure 4.7   (a) The relationship between the size of minnows (arbitrary size classes) and the number of signs of parasitic infection observed on them. $r_s = 0.74$, $n = 11$, $p < 0.02$. (b) The relationship between the size of minnows (arbitrary size classes) and the mean number of signs of parasitic infection observed on them. $r_s = 0.74$, $n = 11$, $p < 0.02$. Bars represent standard errors.

calculating a regression line have been given earlier. You may sometimes come across regression plots which show confidence limits as curved lines above and below the regression line itself. However, we shall not be dealing with these here. For further information, see Sokal and Rohlf (1995).

As with correlation, data can be presented as independent points or, where replicated for particular x-values, as means or medians. Once again, where means or medians are presented, significance testing and the fitting of the line are still done using the individual data points, not the summary statistics. Figure 4.8 presents a regression of the effect of additional food during the breeding season on the number of young moorhens (*Gallinula chloropus*) surviving into their first winter in three study populations. Five different quantities of food were used and the three populations received them in a different order over a five-year experimental period. In Fig. 4.8a, the numbers for each population are presented separately; in Fig. 4.8b they are presented as means (± s.e.) across the three study populations.

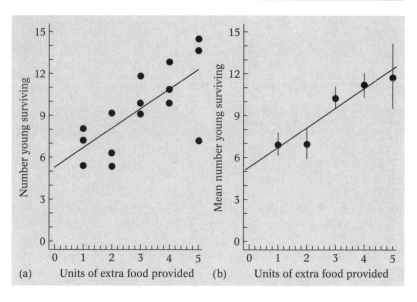

Figure 4.8 (a) The number of chicks surviving to their first winter in relation to the number of units of extra food provided during the breeding season in three populations of moorhen. $F = 13.27$, d.f. $= 1,13$, $p < 0.01$. (b) The mean number of chicks surviving to their first winter in relation to the number of units of extra food provided during the breeding season in three populations of moorhen. $F = 13.27$, d.f. $= 1,13$, $p < 0.01$. Bars represent standard errors.

4.2   Presenting results in the text

So far in this section, we have assumed that results will be presented as figures or tables. Figures and tables, however, take up a lot of space in a report and may not be justified if the result is relatively minor or there is a strict limit on the length of the report. In such cases, analyses can instead be summarized in parentheses in the text of the 'Results' section (*see later*). The usual form for a difference analysis is to quote the summary statistics, test statistic, sample size or degrees of freedom and p-value. Thus the information in Table 4.1a could easily be presented in the text as:

> Treatment with gibberellin resulted in a significant increase in growth compared with non-treated controls (mean ($\pm$ s.e.) height of treated plants $= 14.75 \pm 0.88$ cm, $n = 12$; mean height of controls $= 9.01 \pm 0.63$ cm, $n = 8$; $H = 12.22$, $p < 0.001$)

For a trend, it is usual to quote just the test statistic, sample size or degrees of freedom and the p-value. For instance, the information in Fig. 4.8a could be summarized as follows:

The number of chicks hatching during a breeding season that survived into their first winter increased significantly with the amount of extra food provided within the population ($F = 13.27$, d.f. $= 1,13$, $p < 0.01$)

It is impossible to generalize about when an analysis could be presented in the text rather than in separate figures or tables. Sometimes, as we have said, it is simply a matter of limited space. However, rough guidelines might include the following: (a) difference analyses between only two or three groups; (b) corroborative analysis, supporting a main analysis already presented as a figure or table (for instance, if a main analysis showed a significant correlation between body size and fighting ability, a corroborative analysis might check that body size was not confounded with age and that the correlation could not have arisen because bigger individuals had more experience of fighting); (c) analyses providing background information (e.g. showing a significant sex difference in body size where this is germane to, say, an analysis of the diet preferences of the two sexes).

Units
Whether we are dealing with data in tables, figures or text, it is *essential* that appropriate units of measurement are included and cited with their conventional abbreviations (Box 4.1). Summary statistics are meaningless without them.

Some common units of measurement and their conventional abbreviations	Box 4.1

Length

Kilometre	km
Metre	m
Centimetre	cm
Millimetre	mm

Area

Square kilometre	km^2
Hectare	ha
Square metre	m^2

Square centimetre	cm^2
Square millimetre	mm^2

Volume

Cubic decimetre ($\equiv$ Litre)	dm^3 (l)
Cubic centimetre ($\equiv$ Millilitre)	cm^3 (ml)
Cubic millimetre	mm^3
Microlitre	μl

Weight

Kilogram	kg
Gram	g
Milligram	mg
Nanogram	ng

Time

Million years	My
Years	y
Hours	h
Minutes	min
Seconds	s

4.3   Writing reports

Just as figures and tables of data should be presented properly to ensure they are effective, so care must be taken in the text of a report. In the scientific community, reports of experiments and observations are usually published in the form of papers in professional journals, or sometimes as chapters in specialist books. In all cases, however, the aim is both to communicate the findings of a piece of research and provide the information necessary for someone else to repeat the work and check out the results for themselves. Both these elements are crucial and, as a result, scientific papers are usually refereed by other people in the same field to make sure they come up to scratch before being published. Not surprisingly, a more or less standard format for reports has emerged which divides the textual information into well-recognized sections that researchers expect to see and know how to refer to to find out about different aspects of the work. Learning to use this format properly is one of the most important goals of any basic scientific training. To finish off, therefore, we shall discuss the general structure of a report and what should and should not go in each of its sections; then we shall develop a full report, incorporating our various points about text and data presentation, from some of our main example observations.

4.3.1   The sections of a report

There are five principal sections in a report of experimental or observational work: *Introduction, Methods, Results, Discussion* and *References*. Sometimes it is helpful to have some small additional sections such as *Abstract, Conclusions* and *Appendices*, but we shall deal with these later.

Introduction

The Introduction should set the scene for all that follows. Its principal objective is to set out: (a) the *background* to the study, which means any theoretical or previous experimental/observational work that led to the hypotheses under test. Background information is thus likely to include references to previously published work and sometimes a critical review of competing ideas or interpretations. It might also include discussion about the timing (seasonal, diel, etc.) of experiments/observations and, in the case of fieldwork, the reasons for choosing a particular study site; (b) a clear statement of the hypotheses and predictions that are being tested; and (c) the rationale of the study, i.e. how its design allows the specified predictions to be tested and alternatives to be excluded. The Introduction should thus give the reader a clear idea as to why the study was carried out and what it aimed to investigate. The following is a brief example:

Reptiles are ectotherms and thus obtain most of the heat used to maintain body temperature from the external environment (e.g. Davies, 1979). Rattlesnakes (*Crotalus* spp.) do this by basking in the sun or seeking warm surfaces on which to lie (Bush, 1971).

An increased incidence of snake bites in the state over the past two years has been attributed to a number of construction projects that have incidentally provided rattlesnakes with concrete or tarmac surfaces on which to bask (North, 1989). The aim of this investigation was to study the effect of the construction projects on basking patterns among rattlesnakes to see whether these might increase the exposure of people to snakes and thus their risk of being bitten. The study tests two hypotheses: (a) concrete and tarmac surfaces are preferred basking substrates for rattlesnakes and (b) such surfaces result in a higher than average density of snakes near humans.

When the reader moves on to the Methods and Results sections, they will then appreciate why things were done the way they were. Reading Methods or Results sections without adequate introductory information can be a frustrating and often fruitless business since the design of an experiment or observation usually makes sense only in the context of its rationale. As we shall see below, it is sometimes appropriate to include background material in the Discussion section, but in this case it should be to help develop an interpretation or conclusion, not an afterthought about information relevant to the investigation as a whole; if it is the latter, it should be in the Introduction.

Methods (or Materials and Methods)

The Methods section is perhaps the most straightforward. Nevertheless, there are some important points to bear in mind. Chief among them is providing enough detail for someone else to be able to repeat what you did *exactly*. Clearly, the precise detail in each case will depend on the investigation, but points that need attention are likely to include the following.

Experimental/observational organisms or preparations.   The species, strain, number of individuals used, growth or housing conditions and husbandry, age and sex, etc. for organisms; the derivation and preparation and maintenance techniques, etc. for preparations (e.g. cell cultures, histological preparations, pathogen inoculations).

Specialized equipment.   Details (make, model, relevant technical specifications, etc.) of any special equipment used. This usually means things like tape or video recorders, specialist computing equipment, spectrometers, oscilloscopes, automatic data-loggers, optical equipment such as telescopes, binoculars or specialized microscopes, centrifuges, respirometers, specially constructed equipment such as partitioned aquaria, choice chambers, etc. Run-of-the-mill laboratory equipment like glassware, balances, hotplates and so on don't usually require details, though the dimensions of things like aquaria or other containers used for observation and the running temperature of heating devices, etc. should be given.

Study site (field work).   Where an investigation has taken place in the field, full details of the study site should normally be given. These should include its location (e.g. grid reference) and a description of its relevant features (e.g. size, habitat structure, use by man) and how these were used in the investigation. The date or time of year of the study may also be relevant.

Data collection.   This should include details of all the important decisions that were made about collecting data. Again, it is impossible to generalize, but the following are likely to be important in many investigations: any pretreatment of material before experiments/observations (e.g. isolation of animals, drug treatment, surgical operations, preparation of cell cultures, staining); details of experimental/observational treatments *and controls*; sample sizes and replication; methods of measurement and timing; methods of recording (e.g. check sheets, tape recording, tally counters, etc.); duration and sequencing of experimental/observational periods; details of any computer software used in data collection. Of course, it is important not to go overboard. For instance, it isn't necessary to relate that a check sheet was ticked with a red ball-point pen rather than a black one, but if the pen was used to stimulate aggression in male sticklebacks (which often attack red objects) then it would be relevant to state that a ball-point pen was used and that it was red.

Results

The Results section is in some ways the most difficult to get right. Many students regard it as little more than a dumping ground for all manner of summary and, worse, raw data. Explanation, where it exists at all in such cases, frequently consists of an introductory 'The results are shown in the following figures . . .' and a terminal 'Thus it can be seen . . .'. A glance at any paper in a journal will show that a Results section is much more than this. At the other extreme, explanation within the Results often drifts into speculative interpretation which is more properly the province of the Discussion (*see below*).

A Results section should do two things and *only* two things: first, it should present the data (almost always in some summarized form, of course) necessary to answer the questions posed; and second, it should explain and justify the analytical approach taken so that the reasons for choice of test and modes of data presentation are clear.

The section should thus include a substantial amount of explanatory text, but explanation should be geared solely to the analyses and presentation of data and not the interpretations or conclusions that might be inferred from them. An example might be as follows:

Figure 1 shows that rattlesnakes are significantly more likely to be found on concrete (Figure 1a) and tarmac (Figure 1b) surfaces around dawn and dusk than around midday. Since many of the construction projects in the survey of snake bite incidence have involved highways (Greenbaum *et al.*, 1984), this temporal pattern of basking may result in highest snake/human encounter at times when public conveniences are closed and motorists are forced to relieve themselves at the roadside. Indeed Table 1 shows a strong association for three highways between time of day and number of motorists stopping by the roadside.

It is also important that all the analyses and presentations of data involved in the report appear in the Results section (as figures, tables or in the text) and only in the Results section; no analysis should appear in any other section.

Discussion

The Discussion is the place to comment on whether the results support or refute the hypotheses under test and how they relate to the findings of other studies. The Discussion thus involves interpretation and reasonable speculation, with further details about the material investigated and any corroborative/contradictory/background information as appropriate. As we have said, however, while the Discussion may flesh out, comment, compare and conclude, it should not bring in new analysis. Neither should it develop background information that is more appropriate to the Introduction (*see earlier*). The kind of thing we'd expect might be as follows:

The results suggest that concrete and tarmac surfaces are not favoured for basking by rattlesnakes in comparison with broadly equivalent natural surfaces when relative area is taken into account. One reason for this might be the greater proximity and greater density of cover close to the natural surfaces sampled. Many snakes (Jones, 1981), including rattlesnakes (Wilson, 1976), prefer basking areas within a short escape distance of thick cover. Despite not being preferred by snakes, the greater incidence of bites on concrete and tarmac surfaces can be explained in terms of the greater intensity of use of these surfaces by humans. However, Wilson (1976) has noted that the probability of attack when a snake is encountered increases significantly if there is little surrounding cover. The paucity of cover around the concrete and tarmac samples may thus add to the risk of attack in these environments.

References

Your report should be referenced fully throughout, with references listed chronologically in the text and alphabetically in a headed Reference section at the end. References styles vary enormously between different kinds of report so there is no one accepted format. However, a style used very widely is illustrated below and we suggest using it except where you are explicitly asked to adopt a different style. In this style, references in the text should take the form:

> ... Smith (1979, 1980) and Grant *et al.* (1989) claim that, during a storm, a tree 10 m in height can break wind for over 100 m (*but see* Jones and Green, 1984; Nidley, 1984, 1986)...

In the References list at the end, journal references take the form:

Grant, A. J., Wormhole, P. and Pigwhistle, E. G. (1989) Tree lines and the control of soil erosion. *Int. J. Arbor.* **121**, 42–78.
Jones, A. B. and Green, C. D. (1984) Soil erosion: a critical review of the effect of tree lines. *J. Plant Ecol.* **83**, 101–107.
Smith, E. F. (1979) Planting density and canopy size among deciduous trees. *Arbor. Ecol.* **19**, 27–50.

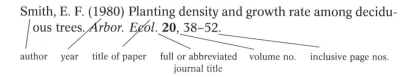

Smith, E. F. (1980) Planting density and growth rate among deciduous trees. *Arbor. Ecol.* **20**, 38–52.

author   year   title of paper   full or abbreviated   volume no.   inclusive page nos.
                                     journal title

for books they take the form:

Nidley, R. (1984) *Deforestation and its impact on national economies*. Hacker Press, London.

title of book in italics   publisher   place of publication

and for chapters in edited volumes the form:

Nidley, R. (1986) Economic growth and deforestation. In *Sustainable economics and world resources*, eds A. B. Jones and C. D. Green, pp. 64–78. Hacker Press, London.

Where more than one source by a particular author (or set of authors) in a particular year is referred to, the sources can be distinguished by using lower case letter suffixes, e.g. (Smith, 1976a, b) indicates that you are referring to two reports by Smith in the year 1976. The order in which you attribute a, b, c, etc. is determined by the order in which you happen to refer to the publications in your report, not the order in which they were published in the relevant year.

Personal observations and personal communications. Although most of the references you will make will be to work by other people, or yourself, that has been published in some form, it is occasionally appropriate to refer to unpublished observations. This usually arises where some previous, but unpublished, observation is germane to an assumption, fact, technique, etc. that you are relying on in your own report. If such observations are your own, they can be referred to in the text as '(personal observation)' or '(pers. obs.)'. If they have been reported to you by someone else, then they can be referred to as, for example, '(P. Smith, personal communication)' or '(P. Smith, pers. comm.)' – note that the name of the person providing the information is given as well.

Other sections of a report
In some cases, there may be additional sections to a report.

Abstract. This is a short summary of the aims and major findings of your investigation. The idea is to provide the reader with a quick overview of what you've done and what is interesting about it so that they can decide whether they want to go into it all in full. Abstracts are generally a half-page paragraph or less and are more usual in longer reports.

Conclusions.   Sometimes, especially where analyses and interpretations are long and involved, it is helpful to highlight the main conclusions in a tail-end section so that the reader finishes with a reminder of the 'take-home' message of the investigation.

Appendix.   Occasionally, certain kinds of information may be incorporated into an Appendix. Such information might include the details of mathematical models or calculations, detailed background arguments, selective raw data or other aspects of the study that potentially might be of importance to readers but which would clutter up and disrupt the main report were they to be included there. Appendices are thus for informative asides that might help some readers but perhaps distract others. It follows, therefore, that appendices should be used selectively, sparingly and for a clear purpose, not as a dumping ground for odds and ends on the grounds that they might just turn out to be useful.

Use of abbreviations.   It is also worth saying something about the use of abbreviations. Many long-winded technical and jargon terms are often abbreviated in reports, papers and books. This is common practice and perfectly acceptable, as long as abbreviations are defined at their first point of use and conventions are adhered to where they exist (some acromyms, for example, are so well-established that people are hard put to recall the full terminology). Thus:

> The hyperstriatum ventrale pars caudalis (HVc) in the forebrain of birds is associated with the production of song. The volume of the HVc also varies with the complexity of song in different species.

and

> To see whether there was any effect of site on the frequency of calling, we carried out a one-way analysis of variance (ANOVA). The results of the ANOVA suggested that site had a profound effect.

present no problem, whereas:

> The HVc in the forebrain of birds is associated with the production of song.

or

> To see whether there was any effect on the frequency of calling, we carried out an ANOVA.

leaves the uninitiated little the wiser.

While abbreviations and acronyms are acceptable, however, they should be used judiciously. Littering text with them is a sure way to destroy its readability and confuse the reader.

4.3.2   Example of a report

Having outlined the general principles of structuring a report, we can finish off by illustrating them more fully in a complete report. The report is one that might arise from some of the experiments we proposed earlier in the main examples, in this case aggression in crickets.

Example report	**The effect of body size on the escalation of aggressive encounters between male field crickets (*Gryllus bimaculatus*)**

Introduction

Fighting is likely to be costly in terms of time and energy expenditure and risk of injury to the individuals involved. We might thus expect natural selection to have favoured mechanisms for reducing the likelihood of costly fights. One way animals could reduce the chance of becoming involved in an escalated fight is to assess their chances of winning or losing against a given opponent before the encounter escalates into all-out fighting. There is now a substantial body of theory (e.g. Parker, 1974; Maynard Smith and Parker, 1976; Enquist *et al.*, 1985) suggesting how assessment mechanisms might evolve and much empirical evidence that animals assess each other during aggressive encounters (e.g. Davies and Halliday, 1978; Clutton-Brock *et al.*, 1979; Austad, 1983). Since the outcome of a fight is likely to be determined by some kind of difference in physical superiority between opponents, features relating to physical superiority might be expected to form the basis for assessment.

Male field crickets (*Gryllus bimaculatus*) compete aggressively for ownership of shelters and access to females (*see* Simmons, 1986). Casual observation of male crickets in a sand-filled arena suggested that body size might be an important determinant of success in fights, with larger males winning more often (pers. obs.). This is borne out by Simmons (1986) who found a similar effect of body size in male *G. bimaculatus*. Observations also showed that aggressive interactions progressed through a well-defined series of escalating stages (*see also* Simmons, 1986) before a fight ensued. One possibility, therefore, is that these escalating stages reflect the acquisition of information about relative body size and interactions progress to the later, more aggressive, stages only when opponents are closely matched in size and the outcome is difficult to predict. This study therefore tests two predictions arising from this hypothesis:

1. large size will confer an advantage in aggressive interactions among male crickets, and

2. interactions will escalate further when opponents are more closely matched in size.

Methods

Four groups of six virgin male crickets were used in the experiment. All males were derived from separate, unrelated stock colonies a week after adult eclosion so each group comprised arbitrarily selected, unfamiliar males on establishment. Crickets were maintained on a 12 h : 12 h light : dark cycle which was shifted by 4 h to allow observation at periods of peak activity (Simmons, 1986). Before establishing a group, the width of each male's pronotum (thorax) was measured at its widest point using Vernier calipers and recorded as an index of the male's body size (the pronotum was chosen because it consists of relatively inflexible cuticle that is unlikely to vary between observations or with handling; adult body size is determined at eclosion so does not change with age). The dorsal surface of the pronotum of each male was then marked with a small spot of coloured enamel paint to allow the observer to identify individuals.

Groups were established in glass arenas (60 × 60 × 30 cm) with 2-cm deep silver sand substrate. Each arena was provided with water-soaked cottonwool in a Petri dish and two to three rodent pellets. No shelters or other defendable objects were provided to avoid bias in the outcome of interactions due to positional advantages. Arenas were maintained under even 60 W white illumination in an ambient room temperature of 25 °C throughout the experiment.

The six males in a group were introduced into their arena simultaneously and allowed to settle for 5 min. They were then observed for 30 min during which time all encounters between males were dictated on to magnetic tape noting: (a) the individuals involved, (b) the individual initiating

Table 1 Degree of escalation increases from Aggressive stridulation to Flip. Each behaviour can thus be ascribed a rank escalation value ranging from 1 (low escalation) to 6 (high escalation)

Behaviour	Description	Escalation ranking
Aggressive stridulation	One or both males stridulate aggressively. This may occur on its own or in conjunction with other aggressive behaviours	1
Antennal lashing	One male whips his opponent with his antennae	2
Mandible spreading	One male spreads his mandibles and displays them to his opponent	3
Lunge	A male rears up and pushes forward, butting the opponent and pushing him backwards	4
Grapple	Males lock mandibles and wrestle	5
Flip	One male throws his opponent aside or on to his back. Re-engagement was rare following a Flip	6

the encounter (the first to perform any of the components of aggressive behaviour – *see below*), (c) the individual that won (decided when one opponent first attempted to retreat) and (d) the components of aggressive behaviour used by each opponent during the encounter. Following Simmons (1986), the aggressive behaviours recognized here are shown in Table 1.

Results

Do larger males tend to win aggressive encounters?   To see whether larger males tended to win more often, the percentage of encounters won by each male in the four groups was plotted against pronotum width (Figure 1). A significant positive trend emerged. Figure 1, however, combined data from all four groups. Did the relationship hold for each group separately? Spearman rank correlation showed a significant relationship in three of the four groups ($r_s = 0.94$, 0.99, 0.97 ($p < 0.05$ in all cases) and 0.66 (ns), $n = 6$ in all groups, one-tailed test).

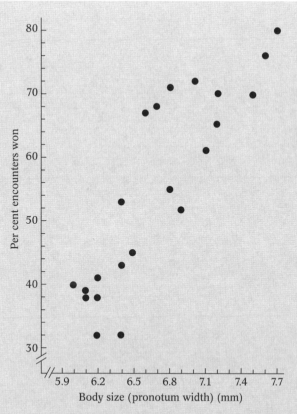

Figure 1 The relationship between the size (pronotum width) of a male and the percentage of encounters won by the male in a group of six. $r_s = 0.81$, $n = 24$, $p < 0.0001$.

If there is a size advantage as suggested by Figure 1, we might expect larger males to initiate more encounters than smaller males since they have more to gain. Figure 2 shows a significant positive correlation between pronotum width and the percentage of the recorded encounters for each male that was initiated (*see* Methods) by that male. As expected, therefore, larger males tended to be the initiator in more of their encounters.

One possibility that arises from Figures 1 and 2 is that the apparent effect of body size was an incidental consequence of the tendency to initiate. There may be an advantage to initiating itself, perhaps because an individual initiates only when its opponent's ability to retaliate is compromised (e.g. it is facing away from its attacker). If the males doing most of the initiating in the groups just happened to be the bigger ones, the

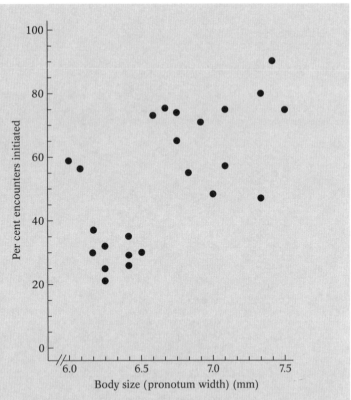

Figure 2 As Figure 1 but for the relationship between male size and the percentage number of encounters initiated by the male. $r_s = 0.63$, $n = 24$, $p < 0.03$.

initiation could underlie the apparent effect of body size on the chances of winning. To test this, the percentage encounters won by each male when he was the initiator was compared with the percentage won when he was not. The analysis showed no significant difference ($U = 82$, n_1, $n_2 = 19$, ns*) between the two conditions.

Does difference in body size affect the degree of escalation in encounters? Figure 3 shows the relationship between the ratio of pronotum width

---

*   Although it is perfectly legitimate to use a Mann–Whitney U-test here, the fact that we are actually comparing data for the two conditions (initiated versus non-initiated encounters) *within* males means we could have used a different sort of two-group difference test (e.g. a Wilcoxon matched-pairs signed ranks test) which takes this into account.

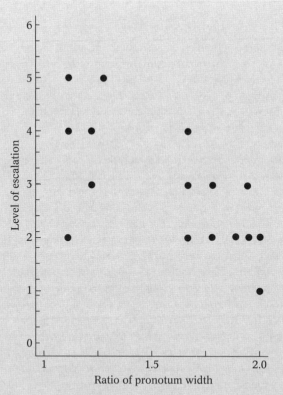

Figure 3 The relationship between ratio of pronotum widths of fighting males and the maximum level of escalation (1–6, *see* Methods) reached in fights. $r_s = -0.71$, $n = 20$, $p < 0.002$.

for pairs of opponents and the degree of escalation of their encounters. A ratio of one indicates equal size and ratios greater than one increasing departure from equality, Degree of escalation is measured as the maximum rank value (1–6, *see* Methods, Table 1) recorded during an encounter. As predicted, the figure shows a significant negative correlation between size ratio and degree of escalation so that escalated encounters were more likely between opponents that were closely matched in size. The trend in Figure 3 is for data across all groups. Does the trend hold within individual males? Correlation analysis for those males (four) that were involved in five or more encounters with opponents of different relative size suggests that it did, although the trends were significant in only two cases ($r_s = -0.96$, $n = 7$, $p < 0.05$; $r_s = -0.72$, $n = 5$, ns; $r_s = -0.76$, $n = 6$, ns; $r_s = -0.99$, $n = 6$, $p < 0.05$, one-tailed test).

Discussion

The results bore out both predictions about the effects of body size on the outcome of aggressive interactions between male field crickets: larger males were more likely to win and escalation was more likely between closely matched opponents. This is consistent with the outcome of fights being largely a matter of physical superiority and with the structuring of interactions into a well-defined series of escalating stages reflecting assessment.

The fact that larger males were more likely to initiate an interaction could mean that the relative size of a potential opponent is assessable in advance of physical interaction. However, it could also reflect a general confidence effect arising from previous wins by larger males (males may initiate according to the simple decision rule 'if I won in the past, I'll probably win this time, so it is worth initiating'). Indeed, Simmons (1986) presents evidence that the number of past wins has a positive influence on the tendency for males to initiate, a result consistent with the confidence effect. Alternatively, initiation could reflect individual recognition, with males picking on those individuals against whom they have won in the past. Since this study did not record encounters independently of the performance of one of the categories of aggressive behaviour, it is not possible to say whether initiations against particular opponents occurred more or less often than expected by chance. Whatever the basis for deciding to initiate, however, there was no evidence that initiation itself conferred an advantage in terms of the outcome.

Although no resources (shelters and females) were available in the arenas, the size advantage in the aggressive interactions recorded here is in keeping with the tendency for larger males to take over shelters and mate successfully with females (Simmons, 1986). While females prefer to mate with males in or near shelters (because these provide good oviposition sites and protection from predators), they will mate with males encountered in open areas (Simmons, 1986). Aggression between males in the absence of shelters or females may thus reflect an advantage to reducing competition should a female happen to be encountered.

References

Austad, S. N. (1983) A game theoretical interpretation of male combat in the bowl and doily spider, *Frontinella pyramitela. Anim. Behav.* **19**, 59–73.

Clutton-Brock, T. H., Albon, S. D., Gibson, R. M. and Guiness, F. E. (1979) The logical stag: adaptive aspects of fighting in red deer (*Cervus elephus* L). *Anim. Behav.* **27**, 211–225.

Davies, N. B. and Halliday, T. R. (1978) Deep croaks and fighting assessment in toads, *Bufo bufo. Nature* **274**, 683–685.

Enquist, M., Plane, E. and Roed, J. (1985) Aggressive communication in fulmars (*Fulmarus glacialis*) competing for food. *Anim. Behav.* **33**, 1107–1120.

Maynard Smith, J. and Parker, G. A. (1976) The logic of asymmetric contests. *Anim. Behav.* **24**, 159–175.

Parker, G. A. (1974) Assessment strategy and the evolution of animal conflicts. *J. Theor. Biol.* **47**, 223–243.

Simmons, L. W. (1986) Inter-male competition and mating success in the field cricket, *Gryllus bimaculatus* (de Geer). *Anim. Behav.* **34**, 567–579.

4.4   Summary

1. Confirmatory analyses are usually presented in summarized form (e.g. summary statistics, scattergrams) as tables or figures or in the text of a report. In all cases, sample sizes (or degrees of freedom), test statistics and probability levels should be quoted. In the case of tables and figures, these can be included within the table or figure itself or within a full, explanatory legend.

2. Results should almost never be presented as raw numerical data because these are difficult for the reader to assimilate. In the exceptional circumstances where the presentation of raw data is helpful, presentation should usually be selective to the points being made and is best incorporated as an Appendix.

3. The axes of figures should be labelled in a way that conveys their meaning clearly and succinctly. Where analyses in different figures are to be compared directly, the axes of the figures should use the same scaling.

4. The legends to tables and figures should provide a complete, self-contained explanation of what they show without the reader's having to search elsewhere for relevant information.

5. Reports of investigations should be structured into clearly defined sections: Introduction, Methods, Results, Discussion, References. Each section has a specific purpose and deals with particular kinds of information. The distinction between them should be strictly maintained. When reports are long or involved it can be helpful to add an Abstract and/or Conclusions section to highlight the main points and take-home messages.

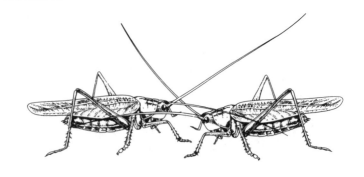

Reference

Sokal, R. R. and Rohlf, F. J. (1995) *Biometry*, 3rd edition, Freeman, San Francisco.

Test finder and help guide

Test finder

NB This key is designed to be used only with the tests described in this book.

A   Looking for a difference or a trend?
    (if unsure go to *Help* 1)

Difference	go to C
Trend	go to B

B   Trying to fit a line and/or predict new values?
    (if unsure go to *Help* 2)

No	use correlation analysis
Yes	use linear regression

C   Data measured at one level of grouping or two?
    (if unsure go to *Help* 3)

One	go to D
Two	go to G

D   Data in *columns* of numbers
    (if unsure go to *Help* 4)

No	go to F
Yes	go to E

E   How many columns?

Two	use *t*-test, Mann–Whitney *U*-test or one-way analysis of variance
More than two	use one-way analysis of variance

F   Data are one row of counts
    (if unsure go to *Help* 5)

Yes	use $1 \times n$ chi-squared test
No	go to G

G   Data are in both *rows* and *columns* (if unsure go to *Help* 6) Data are measurement values (e.g. cm, time, rank, %) or independent counts

with more than one value per row/column cell	use two-way analysis of variance
Data are counts with a single value per row/column cell	use $n \times n$ chi-squared test

Help!

Difference or trend?

■ *Difference* predictions are concerned with some kind of difference between two or more groups of data. The groups could be based on any characteristic that can be used to make a clear-cut distinction, e.g. sex, drug treatment, habitat. Thus a difference might be predicted between the growth rates of men and women, or between the development of disease in rats given drug A versus those given drug B versus those given a placebo.

■ *Trend* predictions are concerned not with differences between mutually exclusive groupings but with the relationship between two more or less continuously distributed measures, e.g. the relationship between the size of a shark and the size of prey it takes, or the relationship between the amount of rainfall in a growing season and the number of apples produced by an apple tree.

Fitting a line and/or predicting a new value?

Fitting a line to a trend by linear regression involves first setting the values of x and then measuring y in relation to these values. Regression then calculates the quantitative relationship between the two measures being related rather than simply seeing whether there is some degree of association between them (which could be achieved using correlation analysis). Knowing the quantitative relationship can be important for two reasons.

1. To get an idea of the magnitude of increase or decrease in one measure with a given increase or decrease in the other, e.g. it wouldn't be surprising to find that bigger sharks ate bigger prey (a correlation coefficient could tell you this), but it might be important to know whether prey size increases at, say, twice the rate of shark size (only regression analysis can give you this kind of information).

2. To predict a value of one measure (e.g. the size of shark that would be dangerous to humans) from a given value of the other (e.g. prey size comparable to a human being). This example prediction could be important in deciding whether the sharks present around a beach constitute a serious threat to bathers, since it wouldn't be ethical to mix sharks and bathers of different sizes to find out by trial and error!

Levels of grouping

Help 3

Many difference predictions are concerned with differences at just *one* level of grouping, e.g. differences in faecal egg counts following treatment of mice with one of four different anthelminthic drugs. Here drug treatment is the only level of grouping in which we are interested. However, if we wished, say, to distinguish between the effects of different drugs on male and female mice, we should be dealing with *two* levels of grouping: drug treatment and sex.

Columns of numbers

Help 4

In these cases, each of the data groupings has more than one data value in it (not necessarily the same number of values in each case) and can be thought of as columns of values under their group headings, for example:

	Group		
	Pesticide A	Pesticide B	Control
% mortality of pest	10	30	0
	5	27	1
	3	50	1
	0	6	0
	1	3	2
	20	3	5

A single row of numbers

Help 5

Here, we are dealing with a single total *count* (e.g. number of organisms, number of events, number of instances on which something met some criterion) under each of the group headings. The numbers cannot be any other kind of measure (e.g. time, length, weight, percentage, ratio). An example would be the number of people questioned who said they preferred a particular soap opera:

	Favourite soap opera			
	Soap 1	Soap 2	Soap 3	Soap 4
Number of people preferring	3	54	71	120

Help 6   Rows and columns

If data have been collected at two levels of grouping, then each data value can be thought of as belonging to both a row and a column (i.e. to one row/column cell) in a table, where rows refer to one level of grouping (say sex – *see* Help 4) and columns to the other (drug treatment – *see* Help 4). If there are several values per row/column cell, as below for the number of individuals dying during a period of observation:

		Treatment	
		Experimental	Control
Sex	Male	3, 4, 8, 12	23, 24, 12, 32
	Female	1, 0, 2, 9	32, 45, 31, 21

then a two-way analysis of variance is appropriate. If there is just a single *count* in each cell, as in the number of male and female fish responding to an experimental or control odour stimulus:

		Treatment	
		Experimental	Control
Sex	Male	27	91
	Female	12	129

then an $n \times n$ chi-squared test is appropriate.

Some self-test questions

(Answers on p. 181)

1. An experimenter recorded the following body lengths of fresh-water shrimps (*Gammarus pulex*) in three different lakes.

Body size (mm)		
Lake 1	Lake 2	Lake 3
9.9	10.5	9.6
8.7	12.1	9.0
9.6	11.2	8.7
10.7	9.7	13.2
8.9	8.7	11.9
8.2	11.1	14.0
7.7	10.7	12.9
8.1	11.8	10.8

   Faunal diversity in the lakes was known ($1 < 2 < 3$) and the experimenter expected shrimps from more diverse lakes to be smaller because of increased interspecific competition. To test this idea he compared body lengths in each pair of lakes (1 versus 2, 2 versus 3 and 1 versus 3) using Mann–Whitney U-tests. Is this an appropriate analysis. If not, what would you do instead?

2. How would you decide between correlation and regression analysis when testing a trend prediction?

3. The following is part of the Discussion section of a report into the effects of temperature and weather on the reproductive rate of aphids on bean plants.

   While the results show a significant increase in the number of aphids produced as temperature rises, there is a possible confounding effect of the age of the host plant and the rate of flow of nutrients. Indeed, there was a stronger significant positive correlation between nutrient flow rate and the number of aphids produced ($r_s = 0.84$, $n = 20$, $p < 0.01$) than between temperature and production (*see* Results).

Do you have any criticisms of the piece?

4. What does the following tell you about the analysis from which it derives?

$$H = 14.1, \qquad \text{d.f.} = 3, \qquad p < 0.01$$

5. An agricultural researcher discovered a significant positive correlation ($r_s = 0.79$, $n = 112$, $p < 0.01$) between daily food intake and the rate of increase in body weight of pigs. What can the researcher conclude from the correlation?

6. A plant physiologist measured the length of the third internode of some experimental plants that had received one of three different hormone treatments. The physiologist calculated the average third internode length for each treatment and for untreated control plants. The data were as follows:

		Treatment		
	Control	Hormone 1	Hormone 2	Hormone 3
Average internode length (mm)	32.3	41.6	38.4	50.2

To see whether there was any significant effect of hormone treatment, the physiologist performed a 1×4 chi-squared test with an expected value of 40.6 in each case and three degrees of freedom. Was this an appropriate test?

7. What do you understand by the terms:
    (a) test statistic,
    (b) ceiling effect,
    (c) statistical significance?

8. The figure shows a significant positive correlation, obtained in the field, between body size in female thargs and the percentage of females in each size class that were pregnant. From this, the observer concluded that male thargs preferred to mate with larger females. Is such a conclusion justified? Give reasons for your answer.

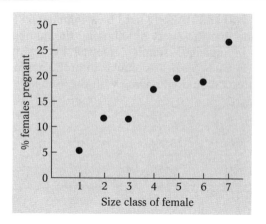

9. Why are significance tests necessary?

10. The following were the results of an experiment to look at the effect of adding an enzyme to its substrate and measuring the rate at which the substrate was split. In the control Treatment A, no enzyme was added; in Treatment B, 10 mg of enzyme was added; in Treatment C 10 mg was added but the reaction was cooled; and in Treatment D, 10 mg was added but the reaction was warmed slightly.

Treatment A	Treatment B	Treatment C	Treatment D
0	20	52	71
1	21	69	92
2	35	100	55
1	15	32	78
0	20		105
0	24		82
			92

What predictions would you make about the outcome of the experiment and how would you analyse the data to test them?

11. A farmer called in an agricultural consultant to help him decide on the best housing conditions (those resulting in the fastest growth) for his pigs. Three types of housing were available (sty + open paddock, crating, and indoor pen). The farmer also kept four different breeds of pig and wanted to know how housing affected the growth rate of each. What analysis might the consultant perform to help the farmer reach a decision?

12. Figures (a) and (b) were used by a commercial forestry company to argue that the effect of felling on the number of bird species living in managed stands (assessed by a single standardized count in each case) was similar in both deciduous and coniferous forest. Would you agree with the company's assessment on this basis?

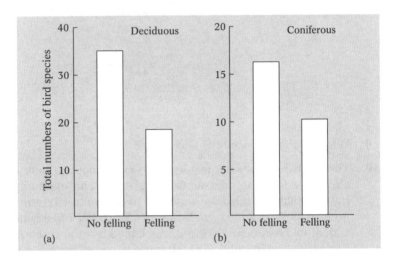

13. What would a *negative* value of the test statistic *r* signify to you?

14. Derive some hypotheses and predictions from the following observational notes.

Sampled some freshwater invertebrates from three different streams using a hand-net. There were more individuals of each species at some sites than others, both within streams and between them. Also some sites had a more or less even distribution of individuals across all species whereas others had a highly biased distribution with some species dominating the community. Some species occurred in all three streams but they tended to be smaller in some streams than others. A number of predatory dragonfly nymphs were recorded but there was never more than one species in any one sample even when more than one existed in a stream. Water quality analyses showed that one stream was badly polluted with effluent from a local factory. This stream and one of the others flowed into the third stream forming a confluence. It was noticed that stones and rocks on the substrate had fewer organisms on or under them in regions of faster flow rate.

15. Is chi-squared used for testing differences or trends?

16. A student in a hall of residence suffered from bed bugs. During the course of a week he was bitten 12 times on his legs, 3 times on his torso, 6 times on his arms and once on his head. Could these data be analysed for site preferences by the bugs? If so, how?

17. An experimenter had counted the number of times kittens showed elements of play behaviour when they were in the presence of their mother or their father and with or without a same-sex sibling. The experimenter had collected ten counts for each condition: (a) mother/sibling, (b) mother/no sibling, (c) father/sibling, (d) father/no sibling, and was trying to decide between a 2×2 chi-squared analysis and a 2×2 two-way analysis of variance. What would you suggest and why?

18. Why do biologists regard a probability of 5 per cent or less as the criterion for significance? Why not be even stricter and use 1 per cent?

19. A fisheries biologist was interested in the maximum size of prey that was acceptable to adult barracuda. To find out what it was, he introduced six adult barracuda into separate tanks and fed them successively larger species of fish (all known to coexist with barracuda in the wild). He then calculated the mean size of the fish that the barracuda last accepted before refusing a fish as a measure of the maximum size they would take. Is this a sensible procedure?

20. A psychologist argued that since males of a species of monkey had larger brains than the females there was less point in trying to teach females complex problem-solving exercises. Any comments?

21. An ecologist studying populations of voles in different woods suspected from a glance at the data that males from some woods had larger adrenal glands than those from others. Unfortunately, the age of the animals also appeared to differ between the woods. How might the ecologist test for a difference in adrenal glands between woods while controlling for the potential confounding effect of age?

22. What do you understand by an 'order effect'?

23. A parasitologist wished to test for the effect of increasing the amount of food supplement to nestling birds on subsequent parasite burdens as adults. The parasitologist intended to carry out a linear regression analysis with amount of food supplement on the x-axis and adult parasite burden on the y-axis. However,

when the distribution of parasite burden data was checked for normality, the following was found:

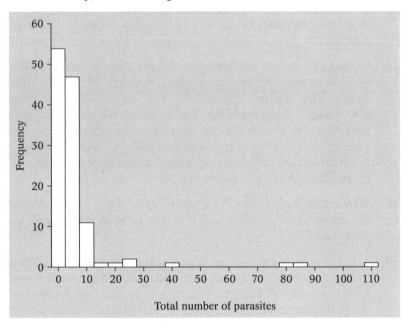

What implications does this have for the regression analysis?

24. What is meant by pseudoreplication?

25. When a botanist compared the frequency of different leaf sizes on a tree to a normal distribution, the significance test comparing the two distributions yielded a test statistic value with an associated probability of 0.0341. Does the botanist's data conform to a normal distribution or not?

Appendix I
Table of confidence limits to the median

Table of nonparametric confidence limits to the median

Sample size (n)	r (for p approx. 95%)
2	–
3	–
4	–
5	–
6	1
7	1
8	1
9	2
10	2
11	2
12	3
13	3
14	3
15	4
16	4
17	5
18	5
19	5
20	6
21	6
22	6
23	7
24	7
25	8
26	8
27	8
28	9
29	9
30	10

r denotes the number of values in from the extremes of the data set that identifies the 95 per cent confidence limits (*see text*). Modified after Colquhoun (1971) *Lectures on biostatistics*, Clarendon Press, Oxford.

Appendix II
Worked examples of significance tests

Examples of tests for a difference between two groups

A.1 Example of a general (two-tailed) *t*-test

Suppose we measure the lengths of nematodes grown in untreated water and in water containing mercuric compounds. The general question is whether the mean lengths of the two groups are different, and hence the null hypothesis H_0 is that there is no difference in the mean lengths of each group. A *t*-test, comparing lengths in the two groups, would go as follows:

Length in untreated water (graticule units)	Length in water treated with mercury
7.2	8.8
7.1	7.5
9.1	7.7
7.2	7.6
7.3	7.4
7.2	6.7
7.5	7.2
mean = 7.514 ± 0.27	mean = 7.557 ± 0.24

1. The sample sizes is 7 in each group, hence $n_1 = n_2 = 7$
2. The variances are calculated as 0.505 and 0.410
3. $P = (7 + 7)/(7 \times 7) = 14/49 = 0.286$
4. $Q = ((6 \times 0.505) + (6 \times 0.410))/(7 + 7 - 2) = (3.03 + 2.46)/12 = 0.458$

5. $R = \sqrt{(0.286 \times 0.458)} = 0.362$

6. The means are 7.514 and 7.557, hence the difference, $S = 7.557 - 7.514 = 0.043$

7. $t = 0.043/0.362 = 0.119$

8. There are $7 + 7 - 2 = 12$ degrees of freedom

9. From Table D in Appendix III, the critical value of the two-tailed t is 2.179, and hence our value is not significant ($p > 0.05$)

10. We conclude that there is no evidence of any difference between the mean values of the two groups ($t = 2.18$, d.f. $= 12$, ns, two-tailed)

A.2 Example of a specific (one-tailed) t-test

Suppose that we measured the weights of five-day old wheat seedlings of normal and genetically modified (GM) seed stocks. From previous results in the literature we were able to make the specific prediction that the mean weight of GM seedlings should be *greater* than normal seedlings. The null hypothesis H_0 is that the mean values do not follow the predicted pattern.

Weight of seedling (g)	
GM stock	Normal stock
10.1	9.5
10.5	9.8
9.8	10.1
10.3	9.7
10.4	9.8
10.4	10.3
10.8	10.0
Mean 10.33 ± 0.12	9.89 ± 0.10

1. The sample size is 7 in each group, hence $n_1 = n_2 = 7$

2. The variances are calculated as 0.099 and 0.071

3. $P = (7 + 7)/(7 \times 7) = 14/49 = 0.286$

4. $Q = ((6 \times 0.099) + (6 \times 0.071))/(7 + 7 - 2) = (0.594 + 0.426)/12 = 0.085$

5. $R = \sqrt{(0.286 \times 0.085)} = 0.156$

6. The mean for the GM seedlings is 10.33, and for the normal seedlings is 9.89. The prediction is that the GM group mean should be greater, hence the difference GM – normal should be positive. Here, $S = 10.33 – 9.89 = 0.44$

7. $t = 0.44/0.156 = 2.82$

8. There are $7 + 7 – 2 = 12$ degrees of freedom

9. From Table D in Appendix III, the critical value of the one-tailed t is 1.782, and hence our value is significant ($p < 0.05$).

10. We conclude that there is evidence that the mean values fall into the predicted pattern: GM seedlings really do weigh more than normal seedlings ($t = 1.78$, d.f. = 12, $p < 0.05$, one-tailed)

A.3   Example of a Mann–Whitney U-test

An ecologist is interested in the effect of microhabitat on the distribution of periwinkles on a rocky shore. Two habitats – a boulder/shingle beach and crevices in a rocky stack – were compared for the prevalence of the commonest periwinkle species measured as the percentage of the total number of all individuals of invertebrate species recorded within quadrat samples. A Mann–Whitney U-test comparing the percentages in replicates of the two habitats is as follows:

% in boulder/shingle	Rank	% in crevices	Rank
40	3.5	60	11
37	1	40	3.5
41	5	55	9
39	2	57	10
43	6	51	8
		63	12
		49	7
	$R_1 = 17.5$		$R_2 = 60.5$

$$U_1 = n_1 \times n_2 + \{[n_1(n_1 + 1)]/2\} – R_1$$

$$n_1 = 5, n_2 = 7$$

$$U_1 = 5 \times 7 + (5 \times 6)/2 – 17.5$$

$$= 35 + 15 – 17.5 = 32.5$$

$$n_1 \times n_2 – U_1 = 5 \times 7 – 32.5 = 2.5$$

Since 2.5 is less than 32.5, $U = 2.5$.

Looking this up in Appendix III, Table B, we see that U is smaller than the critical value of 7 for $p < 0.05$, so we can conclude that there is a significant difference between habitat types in the prevalence of the periwinkle species.

Examples of tests for a difference between two or more groups

B.1   Example of a general parametric one-way analysis of variance

Suppose we have measured the light absorbance of an indicator of a particular biochemical reaction under each of four temperature conditions. We were unable to arrange the experiment so as to have equal numbers of replicates, but we do have several replicates for each group.

Light absorbance of the reaction (arbitrary units)

Group 1 (10 °C)	Group 2 (15 °C)	Group 3 (20 °C)	Group 4 (25 °C)
155	150	170	170
135	156	163	185
145	161	172	175
147	149	158	176
139	162	162	169
152	158	165	174
146	151	166	
	153	164	
	155	153	
	161	173	
		166	
		164	
Mean 145.6 ± 2.6	155.6 ± 1.5	164.7 ± 1.6	174.8 ± 2.3
($n = 7$)	($n = 10$)	($n = 12$)	($n = 6$)

1. The general prediction asks whether there are differences in the mean values of absorbance among temperature groups, hence the null hypothesis is that there are no such differences

2. There are four groups, hence $i = 4$. $N = 35$, with $n_1 = 7$, $n_2 = 10$, $n_3 = 12$ and $n_4 = 6$

3. T, the total sum of squares $= 4238.0$

4. S_i, the sums of squares for each group, are 287.71, 208.4, 346.67 and 162.83

5. The error sum of squares, $SS_{error} = 1005.61$

6. The among-groups sum of squares, $SS_{among} = 4238 - 1005.61 = 3232.39$

7. $d.f._{total} = 34$

8. $d.f._{among} = 3$

9. $d.f._{error} = 34 - 3 = 31$

10. $MS_{among} = 3232.39/3 = 1077.46$

11. $MS_{error} = 1005.61/31 = 32.44$

12. The test statistic, $F = 1077.46/32.44 = 33.2$ for 3,31 degrees of freedom

13. The critical value of $F = 2.91$, d.f. $= 3,31$ and hence the result is significant ($p < 0.05$). In fact, the probability of getting this pattern if the null hypothesis were true is very low indeed, justifying our reporting the fact that $p < 0.001$

14. We conclude that there is evidence for real differences among the mean absorbance values for our temperature groups, and we reject the null hypothesis

15. Results are generally presented in a table, in the following standard format:

Source	SS	d.f.	MS	F	p
Among groups	3232.4	3	1077.5	33.2	<0.001
Error	1005.6	31	32.4		
Total	4238.0	34			

B.2   Example of a specific parametric one-way analysis of variance

We are interested in developing artificial techniques to rear bumblebees for pollinating greenhouse plants. We have measured the weights of bumblebee queens overwintered in the soil under four different conditions. The first group is the normal rearing conditions, but the other groups use combinations of extra factors (stones, leaves and cotton wool). We think any of these factors may improve the weight of the resulting queens. We therefore are interested in two particular sets of contrasts:

- normal versus augmented (all three other groups);

- natural (stones or leaves) versus artificial (cotton wool) additions.

Weights (mg) of bumblebee queens overwintered under different conditions			
Group A (normal)	Group B (stones)	Group C (leaves)	Group D (cotton wool)
310	350	354	376
326	356	360	374
340	358	362	382
338	376	352	362
332	338	366	384
336	342	372	390
344	366	362	361
352	350	344	378
Mean 334.8	354.5	359.0	375.9
s.e. 1.69	1.66	1.17	1.36

1. The general parametric one-way ANOVA gives a test statistic of $F = 18.5$, d.f. $= 3,28$, $p < 0.05$. Among the four groups, there is evidence of significant differences in mean weights, and we therefore reject the null hypothesis of no differences in mean values. In the ANOVA, $SS_{among} = 6862.6$ and $MS_{error} = 123.5$

2. The first contrast is between the normal group and the augmented group. The prediction is that the added components will *increase* the weights of the queens at emergence next spring: this is therefore a specific prediction. Casting it as a contrast:

    $A < (B + C + D)/3$, hence (making the left-hand side > zero)

    $-3.A + 1.B + 1.C + 1.D > 0$

    These coefficients of $-3,1,1,1$ sum to zero, $\sum \lambda_i = 0$, as required

3. The actual mean values of the contrasted groups are 334.8 (normal) and 363.1 (augmented).

    $L = \sum \lambda_i . m_i$
    $= (-3)(334.8) + (1)(354.5) + (1)(359.0) + (1)(375.9) = 85.13$

    This is positive, and hence the mean values lie in the correct direction for the prediction.

$$MS_L = (85.13)^2 . 8/(3^2 + 1^2 + 1^2 + 1^2) = 4830.8$$
$$F = 4830.8/123.5 = 39.11 \text{ and } t = \sqrt{(39.11)} = 6.25$$
$$\text{for } 4 \times 7 = 28 \text{ d.f.}$$

4. Thus for the normal versus augmented contrast, the test statistic is $t = 6.25$ for 28 degrees of freedom. It is a specific prediction, and hence we look up the critical value of one-tailed t-test in Table D of Appendix III, and for 28 d.f. it is 1.7. This contrast is therefore significant: there is evidence that augmenting the environment increases the weight of queen bumblebees over the normal conditions, and we reject the null hypothesis that it does not increase their weight

5. The contrast between adding stones or leaves versus adding cotton wool involves omitting the normal group. Before the experiment, we suspected that augmenting the environment would increase queen weights, but we had no ideas about any differences among different types of augmentation. The prediction is therefore a general one:

    $(B + C)/2 \neq D$, hence (making the left-hand side $\neq$ zero)
    $0.A + 1.B + 1.C - 2.D \neq 0$

    These coefficients sum to zero, as required. They also indicate that this contrast is independent of (orthogonal to) the first contrast, since:

    $$(-3)(0) + (1)(1) + (1)(1) + (1)(-2) = 0$$

6. The actual mean values are 356.8 (stones + leaves) and 375.9 (cotton wool)

    $$L = \Sigma \lambda_i . m_i = -38.25$$

    This is negative, but this does not matter since we are making a general prediction, and don't mind about the direction of the difference.

    $$MS_L = 1950.8, \ F = 1950.8/123.5 = 15.8, \ t = 4.0 \text{ for 21 d.f.}$$

    Since we are testing a general prediction, the critical value of a two-tailed t-test for 21 d.f. is 2.08. This contrast is therefore also significant: there is evidence that using cotton wool improves the weights more than either leaves or stones, and we reject the null hypothesis of no difference

7. We had four groups, and hence were allowed to make three contrasts. Technically what we are doing is 'decomposing the sums of squares', as mentioned before. We can use this fact to see whether any other difference could possibly exist among the groups.

From the one-way ANOVA, $SS_{among} = 6862.6$, and in our two contrasts we have 'taken out' 4830.8 and 1950.8, i.e. 6781.6 (98.8 per cent). Since the test statistic involves dividing by the MS_{error}, i.e. by 123.5, there simply aren't enough sums of squares left to generate any further significant differences

8. Results are generally presented in a table, in the following standard format:

Source	SS	d.f.	MS	F	p
Among groups	6862.6	3	2287.5	18.5	<0.05
Normal versus augmented	4830.8	1	4830.8	$t = 6.3$	<0.001, one-tailed
Stones versus leaves	1950.8	1	1950.8	$t = 4.0$	<0.05
Error	3458.0	28	123.5		
Total	10320.6	31			

B.3   Example of a general and specific nonparametric one-way analysis of variance (modified after Meddis, 1984)

An experimenter measured how much alcohol students in all-female, all-male or mixed halls of residence drank in a month. There are two types of prediction that can be tested:

1. that types of hall differ in their alcohol consumption (general prediction);

2. that all-male halls get through more alcohol than mixed halls and that all-female halls drink the least (specific prediction).

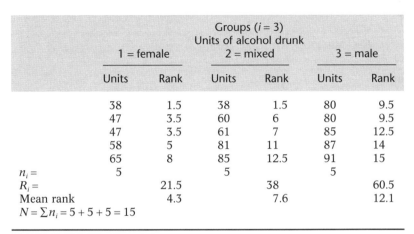

	Groups ($i = 3$) Units of alcohol drunk					
	1 = female		2 = mixed		3 = male	
	Units	Rank	Units	Rank	Units	Rank
	38	1.5	38	1.5	80	9.5
	47	3.5	60	6	80	9.5
	47	3.5	61	7	85	12.5
	58	5	81	11	87	14
	65	8	85	12.5	91	15
$n_i =$	5		5		5	
$R_i =$		21.5		38		60.5
Mean rank		4.3		7.6		12.1
$N = \Sigma n_i = 5 + 5 + 5 = 15$						

The data have been ranked in the table, and then the sums of the ranks for each group have been calculated.

General prediction   *Students in the different types of hall differ in their alcohol consumption.*

Compute

$$H = \frac{12}{N(N+1)} (\Sigma R_i^2/n_i) - 3(N+1)$$

$$H = [12/(15 \times 16)] \times (21.5^2/5 + 38^2/5 + 60.5^2/5) - 3(16)$$

$$= 7.665$$

for two degrees of freedom; the critical value for H (as χ^2) is 5.99 (Appendix III, Table A). Therefore, this value of H is significant at the 5 per cent level.

Specific prediction   *The levels of alcohol consumption follow the order all-female halls < mixed halls < all-male halls.*

The hypothesis is that female < mixed < male, hence the values of λ_i are 1, 2 and 3, since this is the specified rank order of the mean ranks. Thus

$$L = \Sigma \lambda_i R_i = (1)(21.5) + (2)(38) + (3)(60.5) = 279$$

$$\Sigma \lambda_i n_i = (1)(5) + (2)(5) + (3)(5) = 30$$

$$\Sigma n_i (\lambda_i)^2 = (1)(1)(5) + (2)(2)(5) + (3)(3)(5) = 70$$

then

$$E = (N+1)[(\Sigma \lambda_i n_i)/2] = 16(30/2) = 240$$

and

$$V = (N+1)[N \times \Sigma n_i \lambda_i^2 - (\Sigma \lambda_i n_i)^2]/12$$

$$= 16[15(70) - 30^2]/12 = 200$$

$$z = (L - E)/\sqrt{V} = (279 - 240)/\sqrt{200} = 2.76$$

Since this is greater than 1.64 (see Appendix III, Table C), the specified order of mean ranks is significant at the 5 per cent level: therefore $p < 0.05$.

Reference
Meddis, R. (1984) *Statistics using ranks: a unified approach.* Blackwell, Oxford.

Example of a $1 \times N$ chi-squared test

A clinical microbiologist was assaying the effect of four antibiotics on a bacterial culture. To see whether the antibiotics differed in their ability to kill the bacterium, he counted the number of cultures on which clear plaques appeared after drop treatment with each antibiotic and performed a 1×4 chi-squared test assuming equal expected values across antibiotics. The results were as follows:

	Antibiotic treatment			
	A	B	C	D
Observed (O) no. with plaques	21	7	18	30
Expected (E) no.	19	19	19	19

$$\chi^2 = \Sigma \frac{(O - E)^2}{E}$$

$$= \frac{(21 - 19)^2}{19} + \frac{(7 - 19)^2}{19} + \frac{(18 - 19)^2}{19} + \frac{(30 - 19)^2}{19}$$

$$= 0.21 + 7.58 + 0.05 + 6.37$$

$$= 14.21$$

degrees of freedom $= 4 - 1 = 3$

Checking $\chi^2 = 14.21$ for 3 d.f. in Appendix III, Table A shows that it is significant at $p < 0.01$.

Examples of tests for a difference between two groups with two levels of classification

D.1 Example of a general parametric two-way analysis of variance

		Activity of rat liver alcohol dehydrogenase (arbitrary units) Parasitized status		
		Control	Parasitized	
		3.35	4.16	
		4.25	4.59	
		3.66	4.13	
		3.95	4.87	Row mean
	Peas	4.11	4.23	4.18
		4.33	4.38	
		3.75	4.02	
		4.08	4.73	
		4.19	4.46	
		3.86	4.5	
		Mean 3.95	Mean 4.41	
Food type				
		3.38	4.02	
		3.56	4.06	
		3.65	3.89	
		4.03	3.97	
		3.97	4.15	Row mean
	Beans	3.84	3.82	3.79
		3.56	3.73	
		3.43	3.99	
		3.67	3.84	
		3.5	3.74	
		Mean 3.65	Mean 3.92	
	Column means	3.81	4.16	

The above data give the results of an experiment to detect the effects on rat liver alcohol dehydrogenase activity of food type (beans or peas) and parasitism (parasitized or unparasitized). The prediction asks whether there is a difference in enzyme activity between the food types, or between unparasitized and parasitized animals, or an interaction between these two factors. The null hypothesis in each of the cases is that there is no difference. The figures represent arbitrary units of enzymic activity. (Food type and parasitism are regarded as fixed factors.)

1. There are two columns and two rows, with ten replicates per cell

2. The cell, column and row means are shown

3. $SS_{rows} = 1.52$; $SS_{cols} = 1.28$; $SS_{grps} = 2.895$

4. $SS_{int} = SS_{grps} - SS_{cols} - SS_{rows} = 2.89 - 1.52 - 1.28 = 0.09$

5. $SS_{error} = 2.13$

6. $d.f._{cols} = 1$, $d.f._{rows} = 1$, $d.f._{int} = 1 \times 1 = 1$, $d.f._{error} = 2 \times 2 \times (10 - 1) = 36$

7. $MS_{rows} = 1.52/1 = 1.52$
   $MS_{cols} = 1.28/1 = 1.28$
   $MS_{int} = 0.09/1 = 0.09$
   $MS_{error} = 2.13/36 = 0.059$

8. $F_{rows} = 1.52/0.059 = 25.73$
   $F_{cols} = 1.28/0.059 = 21.68$
   $F_{int} = 0.09/0.059 = 1.56$

9. All these F-values have the same degrees of freedom, namely 1 and 36. The threshold value for this is 3.10, and hence two of the three F-values are significant ($p < 0.05$ for food type and for parasitism). We therefore have evidence to reject the null hypothesis of no effect of food type, and of parasitism, but we have no evidence of any interactive effects on liver alcohol dehydrogenase activity

10. Results are generally presented in a table, in the following standard format:

Source	SS	d.f.	MS	F	p
Among rows (food type)	1.52	1	1.52	25.7	<0.001
Among columns (parasitized)	1.28	1	1.28	21.7	<0.001
Interaction					
(food type × parasitized)	0.09	1	0.09	1.6	n.s.
Error	2.13	36	0.06		
Total	5.02	39			

D.2 Example of a specific parametric two-way analysis of variance

		Temperature		
		Cold	Normal	Hot
Motivation	Encouraged	Group A	Group B	Group C
		83	89	85
		83	95	75
		116	75	88
		73	88	91	Row mean
		96	86	85	87.70
		95	90	86
		87	92	96
		80	95	76
		93	89	77
		100	87	80
		Mean 90.6	Mean 88.6	Mean 88.9
	Control	Group D	Group E	Group F
		91	81	75
		91	102	78
		97	95	81
		87	79	85
		96	76	84	Row mean
		84	77	86	84.20
		85	81	77
		79	82	75
		86	85	76
		95	86	74
		Mean 89.1	Mean 84.4	Mean 79.1
	Discouraged	Group G	Group H	Group I
		85	81	70
		88	69	65
		91	90	71
		74	65	75
		76	70	61	Row mean
		81	72	60	73.77
		89	66	55
		92	79	48
		96	67	70
		81	71	55
		Mean 85.4	Mean 73.0	Mean 63.0
	Column means	88.33	82.00	75.33

The above data give the results of an experiment to measure the mathematical competence of students assessed in one of the three conditions (hot, normal or cold) in an environmental chamber. One-third of the students had their motivation deliberately lowered by treating them in a very off-hand manner and generally giving them the impression that they were incompetent. Another one-third were given every encouragement, and the final one-third were treated neutrally (the control). The predictions were:

(i)   high temperature reduces mathematical ability;

(ii) lowered motivation reduces mathematical competence;

(iii) high temperature and lowered motivation interact to reduce mathematical competence much more than either factor on its own

The null hypothesis in each case is that the expected pattern does not occur. (Temperature and motivation are regarded as fixed factors.)

1. The general parametric two-way ANOVA shows that there is a significant effect of temperature ($F = 21.1$, d.f. $= 2,81$, $p < 0.05$) and motivation ($F = 26.2$, d.f. $= 2,81$, $p < 0.001$), and a small but significant interaction ($F = 2.9$, d.f. $= 4,81$, $p < 0.05$). The $MS_{error} = 60.1$

2. *Prediction* 1: high temperature reduces mathematical ability. This is a specific prediction about the three column groups:

$$[(A + D + G) + (B + E + H)]/6 > (C + F + I)/3$$

$$+1.A + 1.B - 2.C + 1.D + 1.E - 2.F + 1.G + 1.H - 2.I > 0$$

Using these coefficients, $L = 59.0$ (positive, and hence in the correct direction), $MS_L = 1933.9$, and t (one-tailed) $= 5.7$ for 81 d.f. ($p < 0.001$). Thus we reject the null hypothesis that high temperature does not reduce mathematical competence

   Prediction 2: lowered motivation reduces mathematical competence. This is a specific prediction about the three row groups:

$$[(A + B + C) + (D + E + F)]/6 > (G + H + I)/3$$

$$+1.A + 1.B + 1.C + 1.D + 1.E + 1.F - 2.G - 2.H - 2.I > 0$$

Using these coefficients, $L = 73.1$ (positive, and hence in the correct direction), $MS_L = 2968.7$, and t (one-tailed) $= 49.4$ for 81 d.f. ($p < 0.001$). Thus we reject the null hypothesis that lowered motivation does not reduce mathematical competence.

Prediction 3: high temperature and lowered motivation will interact to reduce mathematical competence more than expected from their separate effects. This can be tested by ignoring the groups for normal motivation and predicting that the difference between encouraged and discouraged students will get larger at high temperatures. Thus:

$$(C - I) > [(A - G) + (B - H)]/2$$
$$-1.A - 1.B + 2.C\ (+0.D + 0.E + 0.F) + 1.G + 1.H - 2.I > 0$$

Using these coefficients, $L = 20.9$ (positive, and hence in the correct direction), $MS_L = 364.0$, and t (one-tailed) $= 2.46$ for 72 d.f. ($p < 0.05$). Thus we reject the null hypothesis that these groups do not fall into the predicted pattern.

3. Results are generally presented in a table, in the following standard format:

Source	SS	d.f.	MS	F	p
Among rows (motivation)	3 152.4	2	1 576.2	26.2	<0.001
Discouraged versus other two rows	2 968.7	1	2 968.7	$t = 49.4$	<0.001, one-tailed
Among columns (temperature)	2 535.6	2	1 267.8	21.1	<0.001
Hot versus other two columns	1 933.9	1	1 933.9	$t = 5.7$	<0.001, one-tailed
Int. (motivation × temperature)	696.9	4	174.2	2.9	<0.05
Hot, discouraged versus cold, encouraged	364.0	1	364.0	$t = 2.5$	<0.05, one-tailed
Error	4 866.0	81	60.1		
Total	11 250.9	89			

D.3 Example of a general and specific nonparametric two-way analysis of variance (modified after Meddis, 1984)

A team of researchers interested in the consequences of regular running for various measures of health and well-being investigated the joint effects of fat in the diet and running for three or more days a week on levels of blood cholesterol. They measured cholesterol in people who had low or high fat intake and did or did not run on three or more days a week. They then cast their cholesterol measures as follows and carried out a nonparametric two-way analysis of variance.

	Cholesterol scores	
	Did not run	Did run
High fat diet	23, 24, 25, 26, 28 = Group A	18, 18.5, 19.1, 20.1, 20.5 = Group B
Low fat diet	19, 20, 20, 21, 23 = Group C	3, 4, 5, 6, 7 = Group D

Case A: specific tests

1. First we shall assume the team wanted to test some specific predictions (Pr):

   Pr1: *Running will reduce blood cholesterol (i.e. $B + D < A + C$ in the table above).*

   Pr2: *Low fat diet will reduce blood cholesterol (i.e. $C + D < A + B$).*

   Pr3: *The effect of running will be greater when on a low fat diet (i.e. $C - D > A - B$).*

2. Now we rank all the data values irrespective of cell in the table, giving low ranks to low values and averaging ranks for tied values. Thus we arrive at the following table:

	Rank values	
	Did not run	Did run
High fat diet	15.5, 17, 18, 19, 20 = Group A	6, 7, 9, 12, 13 = Group B
Low fat diet	8, 10.5, 10.5, 14, 15.5 = Group C	1, 2, 3, 4, 5 = Group D

3. Work out the sums of the ranks and sample sizes for each cell. These are:

	$i = 1$	$i = 2$	$i = 3$	$i = 4$	Sum
Group (i)	A	B	C	D	
Rank sums (R_i)	89.5	47	58.5	15	210
Sample size (n_i)	5	5	5	5	20
Mean ranks	17.9	9.4	11.7	3	

Check that the ranking is correct by using N, the total sample size, to calculate $N(N + 1)/2$. This should equal the sum of ranks. Here $N(N + 1)/2 = 20(21)/2 = 210$, so the ranking is alright.

4. Test the predictions by obtaining the coefficients, λ_i, for each prediction and then calculating the test statistic z using:

$$z = (L - E)/\sqrt{V}$$

where

$$L = \sum \lambda_i R_i$$

$$E = (N + 1)\sum n_i \lambda_i/2$$

$$V = (N + 1)[N\sum n_i \lambda_i^2 - (\sum n_i \lambda_i)^2]/12$$

Thus for Pr1, the prediction is that $B + D < A + C$, i.e. $+A - B + C - D > 0$, so the coefficients, λ_i, are $+1, -1, +1, -1$.

$$\sum n_i \lambda_i = +5 - 5 + 5 - 5, \qquad \text{sum} = 0$$

$$\sum n_i \lambda_i^2 = +5 + 5 + 5 + 5, \qquad \text{sum} = 20$$

Thus:

$$L = \sum \lambda_i R_i$$
$$= (+1)(89.5) + (-1)(47) + (+1)(58.5) + (-1)(15)$$
$$= 89.5 - 47 + 58.5 - 15$$
$$= 86$$

$$E = (N + 1)(\sum n_i \lambda_i)/2$$
$$= (21)(0)/2$$
$$= 0$$

$$V = (n + 1)[\sum n_i \lambda_i^2 - (\sum n_i \lambda_i)^2]/12$$
$$= (21)[20(20) - (0)^2]/12$$
$$= 700$$

$$\therefore z = (86 - 0)/\sqrt{700}$$
$$= 3.25$$

Looking z up in Appendix III, Table C, we find that a value of 3.25 means there is less than 0.1 per cent ($p < 0.001$) probability that we could have obtained our predicted order by chance. We therefore accept that the fit to our predicted order is significant.

Prediction 2 is that $C + D < A + B$, i.e. that $+A + B - C - D > 0$, with coefficients $+1, +1, -1, -1$ and:

$$\Sigma n_i \lambda_i = +5 + 5 - 5 - 5, \qquad \text{sum} = 0$$

$$\Sigma n_i \lambda_i^2 = +5 + 5 + 5 + 5, \qquad \text{sum} = 20$$

Thus:

$$L = (+1)(89.5) + (+1)(47) + (-1)(58.5) + (-1)(15)$$

$$= 89.5 + 47 - 58.5 - 15$$

$$= 63$$

$$E = (21)(0)/2$$

$$= 0$$

$$V = (21)[20(20) - (0)^2]/12$$

$$= 700$$

$$\therefore z = (63 - 0)/\sqrt{700}$$

$$= 2.38$$

Looking in Table C, Appendix III, again, we see that $z = 2.38$ has an associated probability of occurring by chance of $p < 0.01$. Again, therefore, the fit to our prediction is significant.

For Pr3, $C - D > A - B$, i.e. $-A + B + C - D > 0$, so λ_i are -1, $+1$, $+1$, -1.

$$\Sigma n_i \lambda_i = -5 + 5 + 5 - 5, \qquad \text{sum} = 0$$

$$\Sigma n_i \lambda_i^2 = +5 + 5 + 5 + 5, \qquad \text{sum} = 20$$

$$L = -89.5 + 47 + 58.5 - 15$$

$$= 1$$

$$E = 21(0)/2$$

$$= 0$$

$$V = (21)[20(20) - (0)^2]/12$$

$$= 700$$

$$z = (1 - 0)/\sqrt{700}$$

$$= 0.04$$

This time the associated probability for z is not significant ($p > 0.05$).

Thus the three specific predictions have been tested and the team can conclude that:

1. Running *is* associated with a reduction in blood cholesterol ($z = 3.25$, $p < 0.001$).

2. Low fat diet *is* associated with low blood cholesterol ($z = 2.38$, $p < 0.01$).

3. The effect of running *is not* greater when on a low fat diet ($z = 0.04$, ns).

Case B: general tests

Instead of making specific predictions, the team could have made some general ones. However, the only time it makes sense to do *both* is after *none* of the specific predictions have been supported by the data. If none of the predicted patterns emerges it is then useful to ask whether there are any differences at all between groups.

General predictions can be tested only when there are equal numbers of measures in each cell of the analysis of variance table.

1. The team first formulate their general predictions:

Pr4: *Running will affect blood cholesterol levels.*

Pr5: *The amount of fat in the diet will affect blood cholesterol levels.*

Pr6: *There will be an interaction between running and diet in their effects on blood cholesterol levels.*

2. Testing the predictions:

Pr4: Sum the ranks for exercise groups as $A + C$ and $B + D$ giving:

$$R_1 = 89.5 + 58.5$$

$$= 148, n = 10$$

$$R_2 = 47 + 15$$

$$= 62, n = 10$$

Then

$$H_1 = \left[\frac{12}{N(N+1)} \Sigma \left(\frac{R_i^2}{n_i} \right) - 3(N+1) \right]$$

$$= \left[\frac{12}{20(21)} \left(\frac{148^2}{10} + \frac{62^2}{10} \right) - 3(21) \right]$$

$$= 10.57$$

Looking up H as χ^2 in Appendix III, Table A, shows $H = 10.57$, d.f. $= 1$ to be significant at $p < 0.01$.

Pr5: Sum ranks for A + B and C + D, giving:

$$R_1 = 89.5 + 47$$

$$= 136.5, n = 10$$

$$R_2 = 58.5 + 15$$

$$= 73.5, n = 10$$

$$H_2 = \left[\frac{12}{20(21)} \left(\frac{136.5^2}{10} + \frac{73.5^2}{10} \right) - 3(21) \right]$$

$$= 5.67$$

Table A shows $H = 5.67$, d.f. $= 1$ to have an associated probability of $p < 0.05$.

Pr6: First calculate H for the total number of cells using the four rank sums:

$$H_{tot} = \left[\frac{12}{20(21)} \left(\frac{89.5^2}{5} + \frac{47^2}{5} + \frac{58.5^2}{5} + \frac{15^2}{5} \right) - 3(21) \right]$$

$$= 16.24$$

Then

$$H_3 = H_{tot} - H_1 - H_2$$

$$= 16.24 - 10.57 - 5.67$$

$$= 0$$

Table A shows $H = 0$, d.f. $= 1$ to have an associated probability of $p > 0.05$.

The team can thus conclude that:

1. Running *does* affect blood cholesterol levels ($H = 10.57$, $p < 0.01$).

2. The amount of fat in the diet *does* influence blood cholesterol levels ($H = 5.67$, $p < 0.05$).

3. There is no interaction between running and fat in the diet in determining blood cholesterol levels.

Reference

Meddis, R. (1984) *Statistics using ranks: a unified approach.* Blackwell, Oxford.

Examples of tests for a trend

E.1   Example of a Pearson product-moment correlation analysis

These data bear on the relationship between the weight of the gills and the body weight of crabs.

1. We obviously predict a positive association between gill weight and body weight, and hence this is a specific prediction.

2. The 12 pairs of x–y-values are set out here:

x = body wt (g)	y = gill wt (mg)
14.40	159
15.20	179
11.30	100
2.50	45
22.70	384
14.90	230
1.41	100
15.81	320
4.19	80
15.39	220
17.25	320
9.52	210

3. $S_{xx} = 462.48$

    $S_{yy} = 124\,368.91$

    $S_{yy} = 6561.62$

4. $r = 6561.62/\sqrt{(462.48 \times 124\,368.91)} = 0.87$

5. The one-tailed threshold for ten degrees of freedom is 0.497 (Appendix III, Table E), and hence the result is significant: these data provide evidence of a positive association between body weight and gill weight.

E.2   Example of a Spearman rank correlation analysis

The following data set shows the number of sexual approaches by female chaffinches (y) to males with different plumage brightness scores (x). Brightness scores were compounds of different criteria totalling to a maximum possible value of 60. Is there *any* correlation between male brightness and sexual approaches (general prediction)?

Male	Brightness score	No. approaches by females
1	17	0
2	25	3
3	22	2
4	43	6
5	30	5
6	56	8
7	31	4
8	47	7
9	29	2
10	31	4

Ranking the scores for x and y and calculating the differences (d_i) and square of the difference (d_i^2) between ranks gives:

Rank brightness	Rank no. approaches	d_i	d_i^2
1	1	0	0
3	4	1	1
2	2.5	0.5	0.25
8	8	0	0
5	7	2	4
10	10	0	0
6.5	5.5	1	1
9	9	0	0
4	2.5	1.5	2.25
6.5	5.5	1	1
			$\sum d_i^2 = 9.5$

The Spearman rank correlation coefficient, r_S, can now be calculated as:

$$r_S = 1 - [(6\Sigma d_i^2)/(n^3 - n)]$$
$$= 1 - [6(9.5)/(1000 - 10)]$$
$$= 1 - 0.058$$
$$= 0.942$$

Reference to Appendix III, Table F, shows that for $n = 10$, this value is significant at $p < 0.01$ (two-tailed). We can therefore conclude that approaches are associated with male brightness.

E.3   Example of a linear regression analysis

A biogeographer was interested in the effect of distance from one study site on the number of new species discovered on similar sites. Assuming that the site acted as a source region for the others, the biogeographer predicted that, as distance increased, the number of new species (y-values) would also increase. Five sites were chosen at 10-km intervals (x-values) and the following linear regression analysis performed:

	Distance from site (km)	No. of new species
	10	22
	20	23
	30	25
	40	27
	50	28
Sum	150	125

1.   Calculate x^2 and y^2 and the product xy for every pair.

	x^2	y^2	xy
	100	484	220
	400	529	460
	900	625	750
	1600	729	1080
	2500	784	1400
Sum	5500	3151	3910

2. Calculate

$$S_{xx} = \Sigma x^2 - \frac{(\Sigma x)^2}{n}$$

$$= 5500 - \frac{150^2}{5}$$

$$= 1000$$

3. Calculate

$$S_{yy} = \Sigma y^2 - \frac{(\Sigma y)^2}{n}$$

$$= 3151 - \frac{125^2}{5}$$

$$= 26$$

4. Calculate

$$S_{xy} = \Sigma xy - \frac{(\Sigma x)(\Sigma y)}{n}$$

$$= 3910 - \frac{(150)(125)}{5}$$

$$= 160$$

5. Calculate the slope of the regression line as:

$$b = S_{xy}/S_{xx}$$

$$= 160/1000$$

$$= 0.16$$

6. Calculate the intercept of the line as:

$$a = \bar{y} - b\bar{x}$$

$$= (125/5) - 0.16(150/5)$$

$$= 20.2$$

7. Calculate the standard error to the slope as:

$$s_{y/x}^2 = [1/(n-2)](S_{yy} - S_{xy}^2/S_{xx})$$

$$= (1/(5-2))(26 - 160^2/1000)$$

$$= 0.1333$$

$$\text{s.e.} = \sqrt{(s_{y/x}^2/S_{xx})}$$

$$= \sqrt{(0.1333/1000)}$$

$$= 0.011$$

8. Calculate the test statistic F as follows:

$$\text{Regression sum of squares (RSS)} = (S_{xy})^2/S_{xx}$$

$$= 160^2/1000$$

$$= 25.6$$

$$\text{Regression mean square (RMS)} = \text{RSS}/1$$

$$= 25.6$$

$$\text{Deviation sum of squares (DSS)} = S_{yy} - (S_{xy})^2/S_{xx}$$

$$= 26 - 160^2/1000$$

$$= 0.4$$

$$\text{Deviation mean square (DMS)} = \text{DSS}/(n-2)$$

$$= 0.4/(5-2)$$

$$= 0.1333$$

$$F = \text{RMS}/\text{DMS}$$

$$= 25.6/0.1333$$

$$= 192.0$$

$F = 192$ can now be checked against Appendix III, Table G, for 1 (f_1) and $n - 2 = 3$ (f_2) degrees of freedom, yielding a probability of $p < 0.01$. Our biogeographer can thus conclude that there is a significant positive relationship between distance from the original site and the number of new species.

9. Predicting a new y-value and testing an observed y for departure from prediction.

Having established that there was a general relationship between distance and number of species, the biogeographer wanted to check its predictive value by seeing how closely it could predict the number of new species at a site not included in the original analysis. A new site at 25 km from the original site was therefore chosen.

To predict y for the new x-value (x'), the biogeographer used the linear regression equation $y = a + bx$ incorporating the desired x'-value of 25. The predicted y-value from the equation was thus:

$$y = 20.2 + 0.16(25)$$

$$= 24.2$$

The biogeographer then sampled a site at the new distance and came up with 23 new species, the same as at the 20-km site. To see whether this observed number differed significantly from the prediction of 24.2, the biogeographer calculated the test statistic t as:

$$t = \frac{\text{observed } y - \text{predicted } y}{\text{standard error of the prediction}}$$

where the standard error is calculated as:

$$\text{s.e.} = \sqrt{\{(s^2_{y/x})[1 + 1/n + (x' - \bar{x})/S_{xx}]\}}$$

$$= \sqrt{\{(0.1333)[1 + 1/5 + (25 - 30)^2/1000]\}}$$

$$= \sqrt{0.1633}$$

$$= 0.4041$$

Calculating t for observed and predicted y-values:

$$t = \frac{23 - 24.2}{0.4008}$$

$$= 2.99$$

$$\text{Degrees of freedom} = n - 2$$

$$= 5 - 2$$

$$= 3$$

Checking in Appendix III, Table D shows that $t = 2.99$ for d.f. $= 3$ has an associated probability of $p > 0.05$. Therefore there is no significant difference between the biogeographer's observed y-value and that predicted by the regression equation.

Appendix III
Significance tables

Table A Critical values of chi-squared at different levels of p. To be significant, calculated values must be *greater* than those in the table for the chosen level (0.05, 0.01, 0.001) of p and the appropriate number of degrees of freedom

| | Probability, p | | |
Degrees of freedom	0.05	0.01	0.001
1	3.841	6.635	10.83
2	5.991	9.210	13.82
3	7.815	11.34	16.27
4	9.488	13.28	18.47
5	11.07	15.09	20.51
6	12.59	16.81	22.46
7	14.07	18.48	24.32
8	15.51	20.09	26.13
9	16.92	21.67	27.88
10	18.31	23.21	29.59

Table B   Critical values of Mann–Whitney U at $p = 0.05$. To be significant, values must be *smaller* than those in the table for appropriate sizes of n_1 and n_2

n_2 n_1	3	4	5	6	7	8	9	10	15	20
2	–	–	0	0	0	0	0	0	1	2
3	0	0	1	2	2	2	2	3	5	8
4		1	2	3	4	4	4	5	10	13
5			4	5	6	7	7	8	14	20
6				7	8	10	10	11	19	27
7					11	12	12	14	24	34
8						15	15	17	29	41
9							17	20	34	48
10								20	23	39	55
15								34	39	64	90
20								48	55	90	127

Table C   Probabilities associated with different values of z. The body of the table shows probabilities associated with different values of z. Values of z given to the first decimal place vertically and the second decimal place horizontally. z must therefore exceed 1.64 to be significant at $p < 0.05$

z	.00	.01	.02	.03	.04	.05	.06	.07	.08	.09
1.5	.0668	.0655	.0643	.0630	.0618	.0606	.0594	.0582	.0571	.0559
1.6	.0548	.0537	.0526	.0516	.0505	.0495	.0485	.0475	.0465	.0455
1.7	.0446	.0436	.0427	.0418	.0409	.0401	.0392	.0384	.0375	.0367
1.8	.0359	.0351	.0344	.0336	.0329	.0322	.0314	.0307	.0301	.0294
1.9	.0287	.0281	.0274	.0268	.0262	.0256	.0250	.0244	.0239	.0233
2.0	.0228	.0222	.0217	.0212	.0207	.0202	.0197	.0192	.0188	.0183
2.1	.0179	.0174	.0170	.0166	.0162	.0158	.0154	.0150	.0146	.0143
2.2	.0139	.0136	.0132	.0129	.0125	.0122	.0119	.0116	.0113	.0110
2.3	.0107	.0104	.0102	.0099	.0096	.0094	.0091	.0089	.0087	.0084
2.4	.0082	.0080	.0078	.0075	.0073	.0071	.0069	.0068	.0066	.0064
2.5	.0062	.0060	.0059	.0057	.0055	.0054	.0052	.0051	.0049	.0048
2.6	.0047	.0045	.0044	.0043	.0041	.0040	.0039	.0038	.0037	.0036
2.7	.0035	.0034	.0033	.0032	.0031	.0030	.0029	.0028	.0027	.0026
2.8	.0026	.0025	.0024	.0023	.0023	.0022	.0021	.0021	.0020	.0019
2.9	.0019	.0018	.0018	.0017	.0016	.0016	.0015	.0015	.0014	.0014
3.0	.0013	.0013	.0013	.0012	.0012	.0011	.0011	.0011	.0010	.0010
3.1	.0010	.0009	.0009	.0009	.0008	.0008	.0008	.0008	.0007	.0007
3.2	.0007									
3.3	.0005									
3.4	.0003									
3.5	.00023									
3.6	.00016									
3.7	.00011									
3.8	.00007									
3.9	.00005									
4.0	.00003									

Table D   Critical values of t at different levels of p. To be significant at the appropriate level of probability, values must be *greater* than those in the table for the appropriate degrees of freedom

Degrees of freedom	Probability, p						
	0.05 0.10	0.025 0.05	0.01 0.02	0.005 0.01	0.001 0.002	0.0005 0.001	(one-tailed) (two-tailed)
1	6.314	12.71	31.82	63.66	318.3	636.6	
2	2.920	4.303	6.965	9.925	22.33	31.60	
3	2.353	3.182	4.541	5.841	10.21	12.92	
4	2.132	2.776	3.747	4.604	7.173	8.610	
5	2.015	2.571	3.365	4.032	5.893	6.869	
6	1.942	2.447	3.143	3.707	5.208	5.959	
7	1.895	2.365	2.998	3.499	4.785	5.408	
8	1.860	2.306	2.896	3.355	4.501	5.041	
9	1.833	2.262	2.821	3.250	4.297	4.781	
10	1.812	2.228	2.764	3.169	4.144	4.587	
11	1.796	2.201	2.718	3.106	4.025	4.437	
12	1.782	2.179	2.681	3.055	3.930	4.318	
13	1.771	2.160	2.650	3.012	3.852	4.221	
14	1.761	2.145	2.624	2.977	3.787	4.140	
15	1.753	2.131	2.602	2.947	3.733	4.073	
16	1.746	2.120	2.583	2.921	3.686	4.015	
17	1.740	2.110	2.567	2.898	3.646	3.965	
18	1.734	2.101	2.552	2.878	3.610	3.922	
19	1.729	2.093	2.539	2.861	3.579	3.883	
20	1.725	2.086	2.528	5.845	3.552	3.850	

Table E   Critical values for the Pearson product-moment correlation coefficient r. Values must be *greater* than those in the table to be significant at the indicated level of probability

Degrees of freedom	Probability, p					
	0.05 0.1	0.025 0.05	0.01 0.02	0.005 0.01	0.0005 0.001	(one-tailed) (two-tailed)
1	0.988	0.997	1.000	1.000	1.000	
2	0.900	0.950	0.980	0.990	0.999	
3	0.805	0.878	0.934	0.959	0.991	
4	0.729	0.811	0.882	0.917	0.974	
5	0.669	0.755	0.833	0.875	0.951	
6	0.622	0.707	0.789	0.834	0.925	
7	0.582	0.666	0.750	0.798	0.898	
8	0.549	0.632	0.716	0.765	0.872	
9	0.521	0.602	0.685	0.735	0.847	
10	0.497	0.576	0.658	0.708	0.823	
11	0.476	0.553	0.634	0.684	0.801	
12	0.458	0.532	0.612	0.661	0.780	
13	0.441	0.514	0.592	0.641	0.760	
14	0.426	0.497	0.574	0.623	0.742	
15	0.412	0.482	0.558	0.606	0.725	
16	0.400	0.468	0.543	0.590	0.708	
17	0.389	0.456	0.529	0.575	0.693	
18	0.378	0.444	0.516	0.561	0.679	
19	0.369	0.433	0.503	0.549	0.665	
20	0.360	0.423	0.492	0.537	0.652	
25	0.323	0.381	0.445	0.487	0.597	

Table F Critical values for the Spearman rank correlation coefficient r_S. Values must be *greater* than those in the table to be significant at the indicated level of probability

	Probability, p				
	0.05	0.025	0.01	0.005	(one-tailed)
n	0.10	0.05	0.02	0.01	(two-tailed)
4	1.000				
5	.900	1.000	1.000		
6	.829	.886	.943	1.000	
7	.714	.786	.893	.929	
8	.643	.738	.833	.881	
9	.600	.700	.783	.833	
10	.564	.648	.745	.794	
11	.536	.618	.709	.755	
12	.503	.587	.671	.726	
13	.484	.560	.648	.703	
14	.464	.538	.622	.675	
15	.443	.521	.604	.654	
16	.429	.503	.582	.635	
17	.414	.485	.566	.615	
18	.401	.472	.550	.600	
19	.391	.460	.535	.584	
20	.380	.447	.520	.570	
21	.370	.435	.508	.556	
22	.361	.425	.496	.544	
23	.353	.415	.486	.532	
24	.344	.406	.476	.521	
25	.337	.398	.466	.511	

Table G Critical values of F at $p = 0.05$ (upper) and $p = 0.01$ (P146). To be significant, values must be *greater* than those in the tables for the appropriate degrees of freedom (f_1 and f_2)

$p = 0.05$

f_2 \ f_1	1	2	3	4	5	6	7	8	9	10	12	15	20	30	∞
1	161.4	199.5	215.7	224.6	230.2	234.0	236.8	238.9	240.5	241.9	243.9	245.9	248.0	250.1	254.3
2	18.51	19.00	19.16	19.25	19.30	19.33	19.35	19.37	19.38	19.40	19.41	19.43	19.45	19.46	19.50
3	10.13	9.55	9.28	9.12	9.01	8.94	8.89	8.85	8.81	8.79	8.74	8.70	8.66	8.62	8.53
4	7.71	6.94	6.59	6.39	6.26	6.16	6.09	6.04	6.00	5.96	5.91	5.86	5.80	5.75	5.63
5	6.61	5.79	5.41	5.19	5.05	4.95	4.88	4.82	4.77	4.74	4.68	4.62	4.56	4.50	4.36
6	5.99	5.14	4.76	4.53	4.39	4.28	4.21	4.15	4.10	4.06	4.00	3.94	3.87	3.81	3.67
7	5.99	4.74	4.35	4.12	3.97	3.87	3.79	3.73	3.68	3.64	3.57	3.51	3.44	3.38	3.23
8	5.32	4.46	4.07	3.84	3.69	3.58	3.50	3.44	3.39	3.35	3.28	3.22	3.15	3.08	2.93
9	5.12	4.26	3.86	3.63	3.48	3.37	3.29	3.23	3.18	3.14	3.07	3.01	2.94	2.86	2.71
10	4.96	4.10	3.71	3.48	3.33	3.22	3.14	3.07	3.02	2.98	2.91	2.85	2.77	2.70	2.54
11	4.84	3.98	3.59	3.36	3.20	3.09	3.01	2.95	2.90	2.85	2.79	2.72	2.65	2.57	2.40
12	4.75	3.89	3.49	3.26	3.11	3.00	2.91	2.85	2.80	2.75	2.69	2.62	2.54	2.47	2.30
13	4.67	3.81	3.41	3.18	3.03	2.92	2.83	2.77	2.71	2.67	2.60	2.53	2.46	2.38	2.21
14	4.60	3.74	3.34	3.11	2.96	2.85	2.76	2.70	2.65	2.60	2.53	2.46	2.39	2.31	2.13
15	4.54	3.68	3.29	3.06	2.90	2.79	2.71	2.64	2.59	2.54	2.48	2.40	2.33	2.25	2.07
16	4.49	3.63	3.24	3.01	2.74	2.85	2.74	2.66	2.59	2.54	2.42	2.35	2.28	2.19	2.01
17	4.45	3.59	3.20	2.96	2.81	2.70	2.61	2.55	2.49	2.45	2.38	2.31	2.23	2.15	1.96
18	4.41	3.55	3.16	2.93	2.77	2.66	2.58	2.51	2.46	2.41	2.34	2.27	2.19	2.11	1.92
19	4.38	3.52	3.13	2.90	2.74	2.63	2.54	2.48	2.42	2.38	2.31	2.23	2.16	2.07	1.88
20	4.35	3.49	3.10	2.87	2.71	2.60	2.51	2.45	2.39	2.35	2.28	2.20	2.12	2.04	1.84

$p = 0.01$

f_2 \ f_1	1	2	3	4	5	6	7	8	9	10	12	15	20	30	∞
1	4052	4999	5403	5625	5764	5859	5928	5982	6022	6106	6157	6209	6157	6261	6366
2	98.50	99.00	99.17	99.25	99.30	99.33	99.36	99.37	99.39	99.40	99.42	99.43	99.45	99.47	99.50
3	34.12	30.82	29.46	28.71	28.24	27.91	27.67	27.49	27.35	27.23	27.05	26.87	26.69	26.50	26.13
4	21.20	18.00	16.69	15.98	15.52	15.21	14.98	14.80	14.66	14.55	14.37	14.20	14.02	13.84	13.46
5	16.26	13.27	12.06	11.39	10.97	10.67	10.46	10.29	10.16	10.05	9.89	9.72	9.55	9.38	9.02
6	13.75	10.92	9.78	9.15	8.75	8.47	8.26	8.10	7.98	7.87	7.72	7.56	7.40	7.23	6.88
7	12.25	9.55	8.45	7.85	7.46	7.19	6.99	6.84	6.72	6.62	6.47	6.31	6.16	5.99	5.65
8	11.26	8.65	7.59	7.01	6.63	6.37	6.18	6.03	5.91	5.81	5.67	5.52	5.36	5.20	4.86
9	10.56	8.02	6.99	6.42	6.06	5.80	5.61	5.47	5.35	5.26	5.11	4.96	4.81	4.65	4.31
10	10.04	7.56	6.55	5.99	5.64	5.39	5.20	5.06	4.94	4.85	4.71	4.56	4.41	4.25	3.91
11	9.65	7.21	6.22	5.67	5.32	5.07	4.89	4.74	4.63	4.54	4.40	4.25	4.10	3.94	3.60
12	9.33	6.93	5.95	5.41	5.06	4.82	4.64	4.50	4.39	4.30	4.16	4.01	3.86	3.70	3.36
13	9.07	6.70	5.74	5.21	4.86	4.62	4.44	4.30	4.19	4.10	3.96	3.82	3.66	3.51	3.17
14	8.86	6.51	5.56	5.04	4.69	4.46	4.28	4.14	4.03	3.94	3.80	3.66	3.51	3.35	3.00
15	8.68	6.36	5.42	4.89	4.56	4.32	4.14	4.00	3.89	3.80	3.67	3.52	3.37	3.21	2.87
16	8.53	6.23	5.29	4.77	4.44	4.20	4.03	3.89	3.78	3.69	3.55	3.41	3.26	3.10	2.75
17	8.40	6.11	5.18	4.67	4.34	4.10	3.93	3.79	3.68	3.59	3.46	3.31	3.16	3.00	2.65
18	8.29	6.01	5.09	4.58	4.25	4.01	3.84	3.71	3.60	3.51	3.37	3.23	3.08	2.92	2.57
19	8.18	5.93	5.01	4.50	4.17	3.94	3.77	3.63	3.52	3.43	3.30	3.15	3.00	2.84	2.49
20	8.10	5.85	4.94	4.43	4.10	3.87	3.70	3.56	3.46	3.37	3.23	3.09	2.94	2.78	2.42

Answers to self-test questions

1. No, this is not an appropriate analysis because it involves the multiple use of a two-group test (Mann–Whitney U-test). A significant result becomes more likely by chance the greater the number of two-group comparisons that are made. An appropriate analysis would be a nonparametric one-way analysis of variance which allows a test of the specific prediction that body size will be greatest in Lake 1, intermediate in Lake 2 and least in Lake 3.

2. Regression analysis is appropriate when testing for cause and effect in the relationship between x- and y-values, and the x-values are established in the experiment. The data are measured on some kind of constant interval scale that allows a precise, quantitative relationship to be calculated. Because it depends on establishing a quantitative relationship, predicting new values of one variable from new values of the other also demands regression analysis. In other cases, correlation analysis is necessary. Nonparametric correlation analysis is appropriate for both these kinds of data, and others where x-values are merely measured, but yields only the sign and magnitude of the relationship and makes no assumptions about the cause-and-effect relationship between x- and y-values.

3. The investigator has introduced a new analysis into the Discussion (the correlation between nutrient flow rate and aphid production). The analysis should, of course, be in the Results section.

4. The information tells you that the investigator carried out a nonparametric one-way analysis of variance and that they tested a general prediction, thus using the test statistic H instead of z.

The three degrees of freedom tells you that the analysis compared four groups and the p-value that $H = 14.1$ at d.f. $= 3$ is significant at the 1 per cent level.

5. The researcher can conclude that there is a significant positive association between daily food intake and growth rate, but can't necessarily infer that increased growth rate is caused by greater food intake; it could be that faster growing pigs simply eat more food as a result.

6. No, a chi-squared test is not appropriate here because the data are constant interval measurements and not counts.

7. (a) A test statistic is calculated by a significance test and its value has a known probability of occurring by chance for any given sample size or number of degrees of freedom.

    (b) A ceiling effect occurs where observational or experimental procedures are too undemanding to allow a prediction to be tested; all samples approach the maximum value.

    (c) Statistical significance refers to cases where the probability that a difference or trend as extreme as the one observed could have occurred by chance, if the null hypothesis of no difference or trend is true, is equal to or less than an accepted threshold probability (usually 5 per cent, but sometimes 1 or 10 per cent).

8. No, the observer cannot conclude that male thargs prefer larger females just from this. It could be, for instance, that larger females are simply more mature and thus more likely to conceive. Alternatively, depending on how and when size was measured, pregnant females may be larger precisely because they are pregnant!

9. Significance tests provide a generally accepted, arbitrary yardstick for deciding whether a difference or trend is interesting. The yardstick is the probability that the observed difference or trend could have occurred by chance when there wasn't really such a difference or trend in the population. Random variation in sampling will mean that differences or trends will crop up from time to time just by chance.

10. A reasonable prediction would be that the rate of reaction would be lowest in Treatment A because no enzyme had been added and highest in the warmed enzyme/substrate mixture of Treatment D. By the same rationale, Treatment C should have a lower rate than Treatment B because it is cooler. The predicted order

is thus $A < C < B < D$. A suitable significance test would be a specific form of a nonparametric one-way analysis of variance using z as the test statistic.

11. The consultant could try a two-way analysis of variance. The two levels of grouping would be 'housing condition' and 'breed' with three groups at the first level and four at the second. A dozen or so samples for each combination of housing and breed would be useful, though the same number of samples should be used in each case since general rather than specific predictions are being tested. One thing the consultant should be careful to do is distribute pigs from different families arbitrarily across housing conditions so that any effect of housing on growth rate is not confounded with family-specific growth rates (related pigs might grow at a similar rate because they inherited similar growth characteristics). The analysis would indicate any independent effect of housing and breed and any interaction between the two in influencing growth rate.

12. Although the figures look very similar, they are deceptive because their y-axes are scaled differently. The drop in numbers in felled deciduous forest represents 49 per cent of the number in unfelled forest. In coniferous forest it represents only 38 per cent. In proportional terms, therefore, the impact of felling seems to be greater in deciduous forest. However, the fact that the analysis is based on only single counts means it should act only as an exploratory analysis leading to a properly replicated confirmatory analysis.

13. A negative value of r indicates a negative correlation, i.e. the value of y decreases as that of x increases. The sign of the coefficient is ignored when checking against threshold values of significance.

14. Some predictions can be derived as follows:

(a) *Difference predictions*

Observation   The distribution of individuals across species varied between different sites.

Hypothesis *Differences in the degree of dominance of species within a community vary with the ability of species to compete with others for limited resources.*

Prediction *Dominant species will be those whose individuals win in contests with individuals of other species over the resource they occupy.*

Observation   Individuals of some species are smaller in some streams than in others and some streams are more polluted.

Hypothesis   *Pollution results in reduced body size among some freshwater species.*

Prediction   *The body size of any given species will be smallest in the most polluted stream and largest in the least polluted stream.*

(b) *A trend prediction*

Observation   Fewer organisms were seen attached to the substrate in fast flowing parts of the streams.

Hypothesis   *Flow rate influences the ability of organisms to settle on the substrate.*

Prediction   *If clean substrate is provided in areas of different flow rate, then fewer of the organisms drifting by will settle the faster the flow.*

15.   Differences.

16.   Yes, a 1×4 chi-squared analysis could be carried out, but care would have to be taken in calculating expected values because of the different surface areas of the body sites and their possibly different degrees of vulnerability.

17.   Since the experimenter had collected ten counts for each combination of 'parent' and 'sibling' treatment, a two-way analysis of variance would provide most information. It would allow either general or specific predictions about the effects of 'parent' and 'sibling' treatments and the interaction between them to be tested. The ten values in each case could be totalled and used in a 2×2 chi-squared analysis but this would only test for an overall combined effect of the treatments; much useful information would thus be lost in comparison with the analysis of variance.

18.   The threshold probability of 0.05 is an arbitrarily agreed compromise between the risk of accepting a null hypothesis in error (as would happen by setting the threshold p-value too high) and the risk of rejecting it in error (by setting the threshold too low). For many situations in biology, a threshold of $p = 0.01$ would result in an unnecessary risk of accepting the null hypothesis when it was not true.

19.   Clearly this is not a sensible procedure because it confounds the size of prey with the amount each barracuda has already eaten.

Barracuda may give up at a certain size of fish simply because they are satiated, not because the fish is too big. Also size is confounded with species, some of which may be distasteful or be unpalatable in other ways. Again, therefore, barracuda may reject a fish for reasons other than size.

20. There are at least three logical flaws in this line of reasoning. First, if males are generally larger than females then their brains will be proportionally larger too; any comparison of brain sizes should thus be on a relative scale. Second, should the brains of males turn out to be relatively larger, there is no a priori reason to suppose this will affect learning or any other ability. Third, even if differences in brain size do produce differences in learning, training may overcome any such differences.

21. The analysis requires a test for a difference. There is one level of grouping (different woods), so a one-way analysis of variance of the effect of wood on adrenal gland size would be appropriate. The potentially confounding factor of age may be a problem, but as long as the distributions of adrenal gland size and age are normal, the ecologist could do a parametric analysis of variance and include age as a covariate. The result would then test for an effect of wood controlling for any confounding effect of age.

22. An order effect may occur when there is a systematic confounding of experimental treatment and time. Thus, testing for a difference in the performance of chicks in three different learning environments would suffer from an order effect if birds always encountered environment 1 first, then environment 2 and then environment 3. Any difference between environments may simply be due to being earlier or later in the sequence experienced by the birds (or earlier or later in the day/week/season if individual birds experienced only one of the environments).

23. On the face of it, things don't look good for the parasitologist's regression analysis. The parasite-burden data clearly aren't normal, so violate an important assumption about the y-axis variable in a linear regression. However, all is not lost. It may be possible to normalize the data by performing an appropriate transformation. Indeed, a $\log_{10}$ transformation normalizes the data quite nicely, as the figure below shows (a Kolmogorov–Smirnov one-sample test yielded the outcome $z = 1.113$, $n = 120$, $p = 0.157$, showing the parasite data did not depart significantly from a normal distribution). The parasitologist can therefore carry on with the planned analysis using the transformed data.

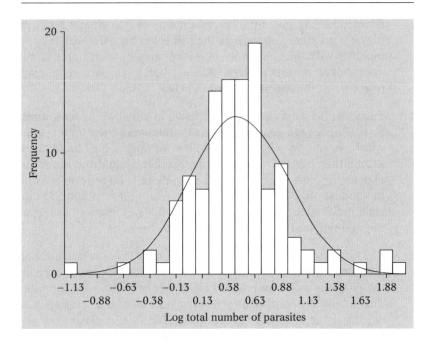

24. Pseudoreplication occurs where there is nonindependence between measures purporting to be independent replicates. An extreme case would be taking repeated measurements of a particular character from the same individual, but more subtle pseudoreplication can arise, for example, when animals from the same cage or litter are treated as independent samples. Animals sharing a cage can have a profound influence on each other behaviourally, physiologically and even morphologically, and those from the same litter obviously share a genetic and experiential background. The number of cages or litters, rather than the number of individuals, thus determines the sample size in any statistical analysis.

25. A bit sneaky this one. Unlike most significance tests of difference, you are looking for a *nonsignificant* outcome when you compare the distribution of data with a normal distribution. If the comparison is not significant, it means your data do not depart from normality more than you would expect by chance. In the example the probability associated with the test outcome was 0.0341, which is less than 0.05. The data thus differ significantly from normal, so the botanist will have to use transformed data if they want to do a parametric significance test, or use a nonparametric test instead.

Index

Quick Test Finder: Tests for Difference

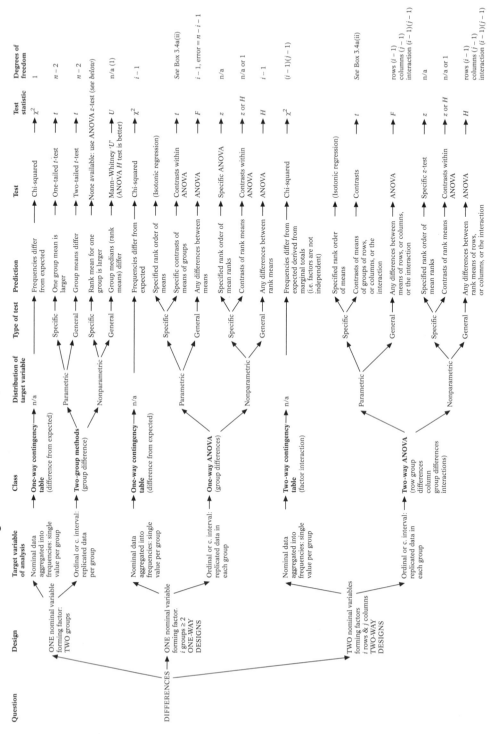

Question	Design	Target variable of analysis	Class	Distribution of target variable	Type of test	Prediction	Test	Test statistic	Degrees of freedom
DIFFERENCES	ONE nominal variable forming factor: TWO groups	Nominal data aggregated into frequencies: single value per group	One-way contingency table (difference from expected)	n/a		Frequencies differ from expected	Chi-squared	χ^2	1
		Ordinal or c. interval: replicated data per group	Two-group methods (group difference)	Parametric	Specific	One group mean is larger	One-tailed t-test	t	$n-2$
					General	Group means differ	Two-tailed t-test	t	$n-2$
				Nonparametric	Specific	Rank mean for one group is larger	None available: use ANOVA z-test (see below)		
					General	Group medians (rank means) differ	Mann–Whitney 'U' (ANOVA H test is better)	U	n/a (1)
	ONE nominal variable forming factor i groups ≥ 2 ONE-WAY DESIGNS	Nominal data aggregated into frequencies: single value per group	One-way contingency table (difference from expected)	n/a		Frequencies differ from expected	Chi-squared	χ^2	$i-1$
		Ordinal or c. interval: replicated data in each group	One-way ANOVA (group differences)	Parametric	Specific	Specified rank order of means	(Isotonic regression)		
						Specific contrasts of means of groups	Contrasts within ANOVA	t	See Box 3.4a(ii)
					General	Any differences between means	ANOVA	F	$i-1$, error $= n-i-1$
				Nonparametric	Specific	Specified rank order of mean ranks	Specific ANOVA	z	n/a
						Contrasts of rank means	Contrasts within ANOVA	z or H	n/a or 1
					General	Any differences between rank means	ANOVA	H	$i-1$
	TWO nominal variables forming factors i rows & j columns TWO-WAY DESIGNS	Nominal data aggregated into frequencies: single value per group	Two-way contingency table (factor interaction)	n/a		Frequencies differ from expected derived from marginal totals (i.e. factors are not independent)	Chi-squared	χ^2	$(i-1)(j-1)$
		Ordinal or c. interval: replicated data in each group	Two-way ANOVA (row group differences column group differences interactions)	Parametric	Specific	Specified rank order of means	(Isotonic regression)		
						Contrasts of means of groups of rows, or columns, or the interaction	Contrasts	t	See Box 3.4a(ii)
					General	Any differences between means of rows, or columns, or the interaction	ANOVA	F	rows $(i-1)$ columns $(j-1)$ interaction $(i-1)(j-1)$
				Nonparametric	Specific	Specified rank order of mean ranks	Specific z-test	z	n/a
						Contrasts of rank means	Contrasts within ANOVA	z or H	n/a or 1
					General	Any differences between rank means of rows, or columns, or the interaction	ANOVA	H	rows $(i-1)$ columns $(j-1)$ interaction $(i-1)(j-1)$